Account Of A Tour In Normandy
Vol. I

By

Dawson Turner

Account Of A Tour In Normandy
Vol. I
by Dawson Turner

ISBN: 978-93-59954-01-1
Published by

DOUBLE 9 BOOKS

2/13-B, Ansari Road
Daryaganj, New Delhi – 110002
info@double9books.com
www.double9books.com
Tel. 011-40042856

ABOUT THE AUTHOR

Dawson Turner FRS FSA was an English banker, botanist, and antiquary who lived from 18 October 1775 to 21 June 1858. He specialized in cryptogam botany and was the father-in-law of botanist William Jackson Hooker and historian Francis Palgrave. Turner was the son of James Turner, the president of the Gurney and Turner's Yarmouth Bank, and Elizabeth Cotman, the only daughter of Yarmouth Mayor John Cotman. He attended North Walsham Grammar School (now Paston College) in Norfolk and Barton Bendish as a botanist's student. He then went to Pembroke College, Cambridge, where his uncle, Rev. Joseph Turner, was the Master. He did not graduate, however, due to his father's grave sickness. He joined his father's bank in 1796. In his spare time after becoming a banker, he became more interested in botany, collecting specimens in the field. Turner promised to assist James Sowerby with specimens in 1794. Turner wrote several books and collaborated with other botanists. He was elected a Fellow of the Royal Society in December 1802. He was named a foreign member of the Royal Swedish Academy of Sciences in 1816. He met Captain George Manby, an amateur artist, inventor, and Yarmouth barrack-master, through his first wife Mary.

CONTENTS

PREFACE

The observations which form the basis of the following letters, were collected during three successive tours in Normandy, in the summers of 1815, 1818, and 1819; but chiefly in the second of these years. Where I have not depended upon my own remarks, I have endeavored, as far as appeared practicable and without tedious minuteness, to quote my authorities for facts; and I believe that I have done so in most instances, except indeed where I have borrowed from the journals of the companions of my tours,— the nearest and dearest of my connections,—or from that of my friend, Mr. Cohen, who, at almost the same time, travelled through a great part of Normandy, pursuing also very similar objects of inquiry. The materials obtained from these sources, it has been impossible to separate from my own; and, interwoven as they are with the rest of the text, it is only in my power to acknowledge, in these general terms, the assistance which I have thus received.—We were proceeding in 1818, to the southern and western districts of Normandy, when a domestic calamity compelled me to return to England. The tour was consequently abridged, and many places of note remained unvisited by us.

My narrative is principally addressed to those readers who find pleasure in the investigation of architectural antiquity. Without the slightest pretensions to the character either of an architect or of an antiquarian, engaged in other avocations and employed in other studies, I am but too conscious of my inability to do justice to the subject. Yet my remarks may at least assist the future traveller, by pointing out such objects as are interesting, either on account of their antiquity or their architectural worth. This information is not to be obtained from the French, who have habitually neglected the investigation of their national monuments. I doubt, however, whether I should have ventured upon publication, if those who have always accompanied me both at home and abroad, had not produced the illustrations which constitute the principal value of my volumes. Of the merits of these illustrations I must not be allowed to speak; but it may be permitted me to observe, that the fine arts afford the only mode of exerting the talents of woman, which does not violate the spirit of the precept which the greatest historian of antiquity has ascribed to the greatest of her heroes—

"Της τε γαρ ὑπαρχουσης φυσεως μη χειροσι γενεσθαι, ὑμιν μεγαλη
ἡ δοξα, χαι δις αν επ᾿ ελαχιστον αρετης περι η ψογου εν τοις αρσεσι
χλεος η."

[English. Not in Original: "Great will be your glory in not falling short
of your natural character; and greatest will be hers who is least talked of
among the men whether for good or for bad." Thucydides' Historiae. (Book
2, Chapter 45, Paragraph 2, Verses 3-5.)]

DAWSON TURNER.

YARMOUTH, *13th August*, 1820.

LETTER I
ARRIVAL AT DIEPPE—SITUATION
AND APPEARANCE OF THE TOWN—
COSTUME OF THE PEOPLE—INHABITANTS
OF THE SUBURB OF POLLET

(Dieppe, June, 1818)

MY DEAR SIR,

You, who were never at sea, can scarcely imagine the pleasure we felt, when, after a passage of unusual length, cooped up with twenty-four other persons in a packet designed only for twelve, and after having experienced every variety that could he afforded by a dead calm, a contrary wind, a brisk gale in our favor, and, finally, by being obliged to lie three hours in a heavy swell off this port, we at last received on board our French pilot, and saw hoisted on the pier the white flag, the signal of ten feet water in the harbor. The general appearance of the coast, near Dieppe, is similar to that which we left at Brighton; but the height of the cliffs, if I am not mistaken, is greater. They vary along the shores of Upper Normandy from one hundred and fifty to seven hundred feet, or even more; the highest lying nearly mid-way between this town and Havre, in the vicinity of Fécamp; and they present an unbroken barrier, of a dazzling white[1], except when they dip into some creek or cove, or open to afford a passage to some river or streamlet. Into one of these, a boat from the opposite shores of Sussex shot past us this afternoon, with the rapidity of lightning. She was a smuggler, and, in spite of the army of Douaniers employed in France, ventured to make the land in the broad face of day, carrying most probably a cargo, composed principally of manufactured goods in cotton and steel. The crew of our vessel, no bad authority in such cases, assured us, that lace is also sent in considerable quantities as a contraband article into France; though, as is well known, much of it likewise comes in the same quality into England, and there are perhaps few of our travellers, who return entirely without it. On the same authority, I am enabled to state, what much surprised me, that the smuggled goods exported from Sussex into Normandy exceed by nearly an hundred fold those received in return.

The first approach to Dieppe is extremely striking. To embark in the evening at Brighton, sleep soundly in the packet, and find yourself, as is commonly the case, early the next morning under the piers of this town, is a transition, which, to a person unused to foreign countries, can scarcely fail to appear otherwise than as a dream; so marked and so entire is the difference between the air of elegance and mutual resemblance in the buildings, of smartness approaching to splendor in the equipages, of fashion in the costume, of the activity of commerce in the movements, and of newness and neatness in every part of the one, contrasted in the other with a strong character of poverty and neglect, with houses as various in their structure as in their materials, with dresses equally dissimilar in point of color, substance, and style, with carriages which seem never to have known the spirit of improvement, and with a general listlessness of manner, the result of indolence, apathy, and want of occupation. With all this, however, the novelty which attends the entrance of the harbor at Dieppe, is not only striking, but interesting. It is not thus at Calais, where half the individuals you meet in the streets are of your own country; where English fashions and manufactures are commonly adopted; and where you hear your native tongue, not only in the hotels, but even the very beggars follow you with, "I say, give me un sou, s'il vous please." But this is not the only advantage which the road by Dieppe from London to Paris possesses, over that by Calais. There is a saving of distance, amounting to twenty miles on the English, and sixty on the French side of the water; the expence is still farther decreased by the yet lower rate of charges at the inns; and, while the ride to the French metropolis by the one route is through a most uninteresting country, with no other objects of curiosity than Amiens, Beauvais, and Abbeville; by the other it passes through a province unrivalled for its fertility and for the beauty of its landscape, and which is allowed by the French themselves to be the garden of the kingdom. Rouen, Vernon, Mantes, and St. Germain, names all more or less connected with English history, successively present themselves to the traveller; and, during the greater part of his journey, his path lies by the side of a noble stream, diversified beyond almost every other by the windings of its channel, and the islands which stud its surface. The only evil to counterbalance the claims of Dieppe is, that the packets do not sail daily, although they profess and actually advertise to that effect; but wait till what they consider a sufficient freight of passengers is assembled, so that, either at Dieppe or Brighton, a person runs the risk of being detained, as has more than once happened to myself, a circumstance that never occurs at Dover. There is still a third point of passage upon our southern coast, and one that has of late been considerably frequented, from Southampton to Havre; but this I never tried, and do not know what it has to recommend it, except to those who are proceeding to Caen or to the western parts of France. The voyage is longer and more uncertain, the distance by land between London and Paris is also greater, nor does it offer equal facilities as to inns and public carriages.

Dieppe is situated on a low tongue of land, but from the sea appears to great advantage; characterized as it is by its old castle, an assemblage of various forms and ages, placed insulated upon an eminence to the west, and by the domes and towers of its churches. The mouth of the harbor is narrow, and inclosed by two long stone piers, on one of which stands an elegant crucifix, raised by the fathers of the mission; to the other has lately been affixed a stone, with an inscription, stating that the Duchess d'Angoulême landed there on her return to her native country; but here is no measure of her foot, no votive pillar, as are to be seen at Calais, to commemorate a similar honor done to the inhabitants by the monarch. A small house on the western pier, is, however, more deserving of notice than either the inscription or the crucifix: it was built by Louis XVIth, for the residence of a sailor, who, by saving the lives of shipwrecked mariners, had deserved well of his sovereign and his country. Its front bears, "A J'n. A'r. Bouzard, pour ses services maritimes;" but there was originally a second inscription in honor of the king, which has been carefully erased. The fury of the revolution could pardon nothing that bore the least relation to royalty; or surely a monument like this, the reward of courage and calculated to inspire only the best of feelings[2], might have been allowed to have remained uninjured. The French are wiser than we are in erecting these public memorials for public virtues: they better understand the art of producing an effect, and they know that such gratifications bestowed upon the living are seldom thrown away. We rarely give them but to the dead. Capt. Manby, to whom above one hundred and thirty shipwrecked mariners are even now indebted for their existence, and whose invention will probably be the means of preservation to thousands, is allowed to live in comparative obscurity; while in France, a mere pilot, for having saved the lives of only eight individuals, had a residence built for him at the public expence, received an immediate gratification of one thousand francs, enjoyed a pension during his life, and, with his name and his exploits, now occupies a conspicuous place in the history of the duchy.

Within the piers, the harbor widens into a stone basin, capable of holding two hundred vessels, and full of water at the flow of the tide; but at the ebb exhibiting little more than a sheet of mud, with a small stream meandering through it. Round the harbor is built the town, which contains above twenty thousand inhabitants, and is singularly picturesque, as well from its situation, backed as it is by the steep cliff to the east, which, instead of terminating here abruptly, takes an inland direction, as from the diversity in the forms and materials of the houses of the quay, some of which are of stone, others of grey flint, more of plaster with their timbers uncovered and painted of different colors, but most of brick, not uncommonly ornamented, with roofs as steep as those of the Thuilleries, and full of projecting lucarnes. This remark, however, applies only to the quay: in its streets, Dieppe is conspicuous among French towns for the uniformity of its buildings. After the bombardment in 1694, when the English, foiled near Brest, wreaked their vengeance upon Dieppe, and reduced the whole to ashes, the town was rebuilt on a regular plan, agreeably to a royal ordinance. Hence this is commonly regarded as one of the handsomest places in France, and you will find it mentioned as such by most authors; but the unfortunate architect who was employed in rebuilding it, got no other reward than general complaints and the nickname of M. Gâteville. The inconveniences arising from the arrangements of the houses which he erected must have been serious; for we find that sixty years afterwards an order of council was procured, allowing the inhabitants to make some alterations that they considered most essential to their comfort. Upon the quay there is occasionally somewhat of the activity of commerce; but elsewhere it is as I have observed before, as well with the people as the buildings. As far as the houses are concerned, a little care and paint would remove their squalid aspect: to an English eye it is singularly offensive; but it cannot possibly be so to the French, among whom it seems almost universal.

To a painter Dieppe must be a source of great delight: the situation, the buildings, the people offer an endless variety; but nothing is more remarkable than the costume of the females of the middle and lower classes, most of whom wear high pyramidal caps, with long lappets entirely concealing their hair, red, blue, or black corsets, large wooden shoes, black stockings, and full scarlet petticoats of the coarsest woollen, pockets of some different die attached to the outside, and not uncommonly the appendage of a key or corkscrew: occasionally too the color of their costume is still farther diversified by a chequered handkerchief and white apron. The young are generally pretty; the old, tanned and ugly; and the transition from youth to age seems instantaneous: labor and poverty have destroyed every intermediate gradation; but, whether young or old, they have all the same good-humored look, and appear generally industrious, though almost incessantly talking. Even on Sundays or feast-days, bonnets are seldom to be seen, but round their necks are suspended large silver or gilt ornaments, usually crosses, while long gold ear-rings drop from either side of their head, and their shoes frequently glitter with paste buckles of an enormous size. Such is the present costume of the females at Dieppe, and throughout the whole Pays de Caux; and in this description, the lover of antiquarian research will easily trace a resemblance to the attire of the women of England, in the XVth and XVIth centuries. As to the cap, which the Cauchoise wears when she appears en grand costume, its very prototype is to be found in Strutt's Ancient Dresses. Decorated with silver before, and with lace streaming behind, it towers on the head of the stiff-necked complacent wearer, whose locks appear beneath, arrayed with statuary precision. Nor is its antiquity solely confined to its form and fashion; for, descending from the great grandmother to the great grand-daughter, it remains as an heir-loom in the family from generation unto generation. In my former visit to Normandy, three years ago, we first saw this head-dress at the theatre at Rouen, and my companion was so struck with it that he made the sketch, of which I send you a copy. The costume of the females of somewhat higher rank is very becoming: they wear muslin caps, opening in front to shew their graceful ringlets, colored gowns, scarlet handkerchiefs, and black aprons.

But nothing connected with the costume or manners of the people at Dieppe is equally interesting as what refers to the inhabitants of the suburb called Pollet; and I will therefore conclude my letter, by extracting from the historian of the place[3] his account of these men, which, though written many years ago, is true in the main even in our days, and it is to be hoped will, in its most important respects, continue so for a length of time to come. "Three-fourths of the natives of this part of the town are fishermen, and not less effectually distinguished from the citizens of Dieppe by their name of Poltese, taken from their place of residence, than by the difference in their dress and language, the simplicity of their manners, and the narrow extent of their acquirements. To the present hour they continue to preserve the same costume as in the XVIth century; wearing trowsers covered with wide short petticoats, which open in the middle to afford room for the legs to move, and woollen waistcoats laced in the front with ribands, and tucked below into the waistband of their trowsers. Over these waistcoats is a close coat, without buttons or fastenings of any kind, which falls so low as to hide their petticoats and extend a foot or more beyond them. These articles of apparel are usually of cloth or serge of a uniform color, and either red or blue; for they interdict every other variation, except that all the seams of their dress are faced with white silk galloon, full an inch in width. To complete the whole, instead of hats, they have on their heads caps of velvet or colored cloth, forming a tout-ensemble of attire, which is evidently ancient, but far from unpicturesque or displeasing. Thus clad, the Poltese, though in the midst of the kingdom, have the appearance of a distinct and foreign colony; whilst, occupied incessantly in fishing, they have remained equally strangers to the civilization and politeness, which the progress of letters during the last two centuries has diffused over France. Nay, scarcely are they acquainted with four hundred words of the French language; and these they pronounce with an idiom exclusively their own, adding to each an oath, by way of epithet; a habit so inveterate with them, that even at confession, at the moment of seeking absolution for the practice, it is no uncommon thing with them to swear they will be guilty of it no more. To balance, however, this defect, their morals are uncorrupted, their fidelity is exemplary, and they are laborious and charitable, and zealous for the honor of their country, in whose cause they often bleed, as well as for their priests, in defence of whom they once threatened to throw the Archbishop of Rouen into the river, and were well nigh executing their threats."

Footnotes:

[1] *The chalk in the cliff, in the immediate vicinity of Dieppe, is divided at intervals of about two feet each by narrow strata of flint, generally horizontal, and composed in some cases of separate nodules, which are not uncommonly split, in others of a continuous compressed mass, about two or three inches thick and of very uncertain extent, but the strata are not regular.*

[2] *Goube Histoire de Normandie, III. p. 188.—In Cadet Gassicourt Lettres sur Normandie, I. p. 68, the story of Bouzard is given still more at length.*

[3] *Histoire de Dieppe, II. p. 56.*

LETTER II
DIEPPE—CASTLE—CHURCHES—HISTORY OF THE PLACE—FEAST OF THE ASSUMPTION

(Dieppe, June, 1818.)

The bombardment of this town, alluded to in my last, was so effectual in its operation, that, excepting the castle and the two churches, the place can boast of little to arrest the attention of the antiquary, or of the curious traveller. These three objects were indeed almost all that escaped the conflagration; and for this they were indebted to their insulated situations, the first on an eminence unconnected with the houses of the place, the other two in their respective cemeteries.

The hill on which the castle stands is steep; and the building, as well from its position, as from its high walls, flanked with towers and bastions, has an imposing appearance. In its general outline it bears a resemblance to the castle of Stirling, but it has not the same claims to attention in an architectural point of view. It is a confused mass of various æras, and its parts are chiefly modern: nor is there any single feature that deserves to be particularized for beauty or singularity; yet, as a whole, a picturesque and pleasing effect results from the very confusion and irregularity of its towers, roofs, and turrets; and this is also enhanced by a row of lofty arches, thrown across a ravine near the entrance, supporting the bridge, and appearing at a distance like the remains of a Roman aqueduct. What seems to be the most ancient part is a high quadrangular tower with lofty pointed pannels in the four walls; and though inferior in antiquity, an observer accustomed only to the English castellated style, is struck by the variety of numerous circular towers with conical roofs, resembling those which flanked the gates of the town. Some of these gates still remain perfect; and one of them, leading to the sea, now serves as a military prison. It was the Sieur des Marêts[4], the first governor of the place, who began this castle shortly after the year 1443, when Louis the XIth, then dauphin, freed Dieppe from the dominion of the English, attacking in person, and carrying by assault, the formidable fortress, constructed by Talbot, in the suburb of Pollet. Of this, not a vestige now remains: the whole was levelled with the ground in 1689; though, at a period of one hundred and twenty years after it was originally taken and dismantled, it had again been made a place of strength by the Huguenots, and had been still further fortified under Henry IVth, in whose reign the present castle was completed; for it was not till this time that permission was given to the inhabitants to add to it a keep. In its perfect state, whilst defended by this keep, and still further protected by copious out-works and bomb-proof casemates, its strength was great; but the period of its power was of short duration; for the then perturbed state of France naturally gave rise to anxiety on the part of the government, lest fortresses should serve as rallying points to the faction of the league; and the castle of Dieppe was consequently left with little more than the semblance of its former greatness.

Of the churches here, that of St. Jaques is considerably the finest building, and is indeed an excellent specimen of what has been called the decorated English style of architecture, the style of this church nearly coinciding in its principal lines with that which prevailed in our own country during the reigns of the second and third Edward. It was begun about the year 1260, but was little advanced at the commencement of the following century; nor were its nineteen chapels, the works of the piety of individuals, completed before 1350. The roof of the choir remained imperfect till ninety years afterwards, whilst that of the transept is as recent as 1628[5]. The most ancient work is discernible in the transepts, but the lines are obscured by later additions. A cloister gallery fronted by delicate mullions runs round the nave and choir, and the extent and arrangement of the exterior would induce a stranger, unacquainted with the history of the building, to suppose that he was entering a conventual or cathedral church. The parts long most generally admired by the French, though they have always been miserable judges of gothic architecture, were the vaulted roof, and the pendants of the Lady-Chapel. The latter were originally ornamented with female figures, representing the Sibyls, made of colored terra cotta, and of such excellent workmanship, that Cardinal Barberini, when he visited this chapel in 1647, declared he had seen nothing of the kind, not even in Italy, superior to them for the beauty and delicacy of their execution; but they are now gone, and, according to Noel[6], were destroyed at the time of the bombardment. The state, however, of the roof does not seem to warrant this observation; and, contrary also to what he says, the pendants between the Lady-Chapel and the choir are still perfect, and serve, together with numerous small canopies in the chapel itself, to give a clear idea of what the whole must have been originally. One of the most elegant of the decorations of the church is a spirally-twisted column, elaborately carved, with a peculiarly fanciful and beautiful capital, placed against a pillar that separates the two south-eastern chapels of the choir. The richest object is a stone-screen to a chantry on the north side, which is divide into several canopies, whose upper part is still full of a profusion of sculpture, though the lower is sadly mutilated. I could not ascertain its history or use; but I do not suppose it is of earlier date than the age of Francis Ist, as the Roman or Italian style is blended with the Gothic arch. The Chapel of the Sepulchre, is not uncommonly pointed out as an object of admiration. There is certainly some, handsome sculpture round the portal; but it is not this for which your admiration is required: you are told that the chapel was made in 1612, at the expence of a traveller, then just returned from Palestine, and that it offers a faithful representation of the Holy Sepulchre itself at Jerusalem; by which if we are to understand that the wretched, grisly, painted, wooden figures of the three Maries, and other holy women and holy men, assembled round a disgusting representation of the dead Saviour, have their prototype in Judea, I can only add I am sorry for it: for my own part, putting aside all question of the propriety or effect of symbolical worship, and meaning nothing offensive to the Romish faith, I must be allowed to say that most assuredly I can conceive nothing less qualified to excite feelings of devotion, or more certain to awaken contempt and loathing, than the images of this description, the tinselled virgins, and the wretched daubs, nick-named paintings, which abound in the churches of Picardy and Normandy, the only catholic provinces which I have yet visited; so that, if the taste of the inhabitants is to be estimated by the decoration of the religious buildings, this faculty must be rated very low indeed. The exterior of the church is as richly ornamented as the inside; and not a buttress, arch, or canopy is without the remains of crumbled carving, worn by time, or disfigured by the ruder hand of calvinistic or revolutionary violence. Tradition refers the erection of this edifice to the English. From the certainty with which a date may be assigned to almost every part, it is very interesting to the lover of architecture. The Lady-Chapel is also perhaps one of the last specimens of Gothic art, but still very pure, except in some of

the smaller ornaments, such, as the niches in the tabernacles, which end in escalop shells.

The other church is dedicated to St. Remi, and is a building of the XVIIth century; though, judging from some of its pillars, it would be pronounced considerably more ancient. Those of the transept and of the central tower are lofty and clustered, and of extraordinary thickness; the rest are circular and plain, and not very unlike the columns of our earliest Norman or Saxon churches, though of greater proportionate altitude. The capitals of those in the choir are singularly capricious, with figures, scrolls, &c.; but it is the capriciousness of the gothic verging into Grecian, not of the Norman. On the pendants of the nave are painted various ornaments, each accompanied by a mitre. The eastern has only a mitre and cross, with the date 1669; the western the same, with 1666; denoting the æra of the edifice, which was scarcely finished, when a bomb, in 1694, destroyed the roof of the choir, and this remains to the present hour incomplete. The most remarkable object in the church is a bénitier of coarse red granite, on whose basin is an inscription, to me illegible. The annexed sketches will give you some idea of it:

In the letters one looks naturally for a date: the figures that alternate with them are probably mitres, and, like those on the roof, indicate the supreme jurisdiction of the Archbishop of Rouen in the place.

Dieppe itself is, by its own historians[7], said to boast an origin as early as the days of Charlemagne[8], who is reported to have built a fortress on the scite of the present town, and to have called it Bertheville, in honor of the Berthas, his mother and his daughter. Bertheville was one of the first places taken by the Normans, by whom the appellation was changed to Dyppe or Dieppe, a word which in their language is said to signify a good anchorage. Other writers[9], however, treat the whole of the early chronicle of Dieppe as a fiction, and maintain, that even at the beginning of the XIth century the town had no existence, and the place was only known as the port of Arques, within whose territory it was comprehended; nor was it till the end of the same century that the inhabitants of Arques were, partly from the convenience of the fisheries, and partly from the advantages of the salt trade, induced to form this settlement. Whatever date may be assigned to the foundation of Dieppe, it is frequently contended that William the Conqueror embarked here for the invasion of England, and it seems undoubted that he sailed hence for his new kingdom in the next year, agreeably to the following passage from Ordericus Vitalis, (p. 509) by which you will observe, that the river had at that time the same name as the town, "Deinde sextâ nocte Decembris ad ostium amnis Deppæ ultra oppidtim Archas accessit, primâque vigiliâ gelidæ noctis Austro vela dedit, et mane portum oppositi littoris, (quem Vvicenesium vocitant) prospero cursu arripuit." In 1188, our Henry II built a castle upon the same hill on which the present fortress stands. This strong hold, however, afforded little protection; for we find that, in 1195, Philip Augustus of France, entering Normandy with an hostile army, laid siege to Dieppe, and set fire not only to the town, but also to the shipping in the harbor. Two years subsequently to this event, Dieppe ceased to form a part of the demesne of the Sovereign of the Duchy. Richard the Ist had given great offence to Walter, Archbishop of Rouen, by persisting in the erection of Château Gaillard, in the vicinity of Andelys, which belonged to the archbishop in right of his see; and though our lion-hearted monarch was not appalled either by the papal interdict or by the showers of blood that fell upon his workmen, yet at length he thought it advisable to purchase at once the forgiveness of the prelate and the secular seignory of Andelys, by surrendering to him, as an equivalent, the towns and lordships of Dieppe and Louviers, the land and forest of Alihermont, the land and lordship of Bouteilles, and the mills of Rouen. This exchange was regarded as so great a subject of triumph to the archbishop, that he caused the memory of it to be perpetuated by inscriptions upon crosses in various parts of Rouen, some of which remained as late as 1610, when Taillepied wrote his Recueil des Antiquitéz et Singularitéz de la Ville de Rouen. The following lines are given as one of these inscriptions in the Gallia Christiana[10]:

"Vicisti, Galtere, tui sunt signa triumphi

Deppa, Locoveris, Alacris-mons, Butila, molta,

Deppa maris portus, Alacris-mons locus amœnus,

Villa Locoveris, rus Butila, molta per urbem.

Hactenus hæc Regis Richardi jura fuere;

Hæc rex sancivit, hæc papa, tibique tuere[11]."

Nor was this the only memorial of the fact; for the advantages of the exchange were so generally recognized, that the name of Walter became proverbial; and to this day it is said in Normandy of a man who over-reaches another, "c'est un fin Gautier." It might be inferred from the terms of the bargain in which Dieppe merely appears as one of the items of the account, that it was then a place of little consequence; yet, one of the old chroniclers speaks of it at the time it was taken by the French under Philip Augustus, as

"portus famâ celeberrimus atque

Villa potens opibus."

These historians, however, of former days are not always the most accurate; but from this period the annals of the place are preserved, and at certain epochs it is far from unimportant in French history: as, when Talbot raised in 1442 the fortress called the Bastille, a defence so strong and in so well-chosen a situation, that even Vauban honored its memory by lamenting its destruction; when the inhabitants fought with the Flemings in the channel, in 1555; when Henry IVth, with an army of less than four thousand men, fled hither in 1589, as to his last place of refuge, winning the hearts of the people by his frank address:—"Mes amis, point de cérémonie, je ne demande que vos cœurs, bon pain, bon vin, et bon visage d'hôtes;" and when, as I have already mentioned, the town sustained from our fleet a bombardment of three days' duration, and was reduced by it to ashes.

For the excellence of its sailors, Dieppe has at all times been renowned: no less an authority than the President de Thou has pronounced them to be men, "penes quos præcipua rei nauticæ gloria semper fuit;" and they have proved their claims to this encomium, not only by having supplied to the navy of France the celebrated Abraham Du Quesne, the successful rival of the great Ruyter, but still more so by having taken the lead in expeditions to Florida[12]; by having established a colony for the promotion of the fur trade in Canada, if indeed they were not the original discoverers of that country; and by having been the first Christians who ever made a settlement on the coast of Senegal. This last-mentioned event took place, according to French writers, at as early a period as the XIVth century; and, though the establishment was not of long duration, its effects have been permanent; for it is owing to the consignments of ivory then made to Dieppe, that many of the inhabitants were induced to become workers in that substance; a trade which they preserve to the present time, and carry the art to such perfection that they have few rivals. This and the making of lace are the principal employments of such of the natives as are not engaged in the fishery. In the earlier ages of the Duchy, the inhabitants of the Pays de Caux found a more effectual and important employment in the salt-works which were then very numerous on the coast, but which have long since been suffered to fall into decay. Ancient charters, recorded in the Neustria Pia, trace these works on the coast of Dieppe, and at Bouteilles on the right of the valley of

Arques, to as remote a period as 1027; and they at the same time prove the existence of a canal between Dieppe and Bouteilles, by which in 1390 vessels loaded with salt were wont to pass. But here, as in England, such works have been abandoned, from the greater facility of communication between distant places, and of obtaining salt by other means.

At present the only manufacture on the beach is that of kelp, for which a large quantity of the coarser sea-weeds is burned; but the fisheries, which are not carried on with equal energy in any other port of France, are the chief support of the place. The sailors of Dieppe were not confined to their own seas; for they used to pursue the cod fishery on the coast of Newfoundland with considerable success. The herring fishery however was a greater staple; and previously to the revolution, when alone a just estimate could be formed of such matters, the quantity of herrings caught by the boats belonging to Dieppe averaged more than eight thousand lasts a year, and realized above £100,000. This fishery is said to have been established here as early as the XIth century[13]. From sixty to eighty boats, each of about thirty tons and carrying fifteen men, were annually sent to the eastern coast of England about the end of August; and then, again, in the middle of October nearly double the quantity of vessels, but of a smaller size, were engaged in the same pursuit on their own shores, where the fish by this time repair. The mackerel fishery was an object of scarcely less importance than that of herrings, producing in general about one hundred and seventy thousand barrels annually. Great quantities of these fish are eaten salted and dried, in which state they afford a general article of food among the lower classes in Normandy. Surely this would be deserving of the attention and imitation of our merchants at home. During the war with England this branch of trade necessarily suffered; but Napoléon did every thing in his power to assist the town, by giving it peculiar advantages as to ships sailing under licences. He succeeded in his views; and, thus patronized, Dieppe flourished exceedingly, and the gains brought in by the privateers connected with the port, added not a little to its prosperity. Hence to this hour the inhabitants regret the peace, although the town cannot fail to be benefitted by the fresh impulse given to the fisheries, and the quantity of money circulated by the travellers who are continually passing. Napoléon intended also to bestow an additional boon upon the place. A canal had been projected many years ago, in the time of the Maréchal de Vauban, and was to have extended to Pontoise, through the fertile districts of Gournay and Neufchâtel, and to have communicated by different branches with the Seine and Oise. This plan, which had been forgotten during so many reigns, Napoléon determined to carry into effect, and the excavations were actually begun under his orders. But the events which succeeded his

Russian campaign put a stop to this, as to all similar labors: the plan is now, however, again in agitation, and, if performed, Dieppe will soon become one of the most important ports in France.

By the revolution Dieppe was emancipated from the dominion of the Archbishop of Rouen, who, by virtue of the cession made by Richard Cœur de Lion, exercised a despotic sway, even until the dissolution of the ancien régime. His privileges were oppressive, and he had and made use of the right of imposing a variety of taxes, which extended even to the articles of provision imported either by land or sea. Yet it must be admitted that the progress of civilization had previously done much towards the removal of the most obnoxious of the abuses. The times, happily, no longer existed, when, as in the XIIth century, the prelate, with a degree of indecency scarcely to be credited, especially under an ecclesiastical government, did not scruple to convert the wages of sin into a source of revenue, as scandalous in its nature as it must have been contemptible in its amount, by exacting from every prostitute a weekly tax of a farthing, for liberty to exercise her profession[14].

Many uncouth and frivolous ecclesiastical rites and ceremonies of the middle ages, which good sense had banished from most other parts of France, where they once were common, still lingered in the archbishop's seignory. Thus, at no very remote period, it was customary on the Feast of Pentecost to cast burning flakes of tow from the vaulting of the church; this stage-trick being considered as a representation of the descent of the fiery tongues. The Virgin, the great idol of popery, was honored by a pageant, which was celebrated with extraordinary splendor; and as I must initiate you in the mysteries of Catholicism, I think you will be well pleased to receive a detailed account of it. The ceremony I consider as curiously illustrative of the manners of the rulers, of the ruled, and of the times; and I will only add, by way of preface, that it was instituted by the governor, Des Marêts, in 1443, in honor of the final expulsion of the English, and that he himself consented to be the first master of the Guild of the Assumption, under whose auspices and direction it was conducted.—About Midsummer the principal inhabitants used to assemble at the Hôtel de Ville, and there they selected the girl of the most exemplary character, to represent the Virgin Mary, and with her six other young women, to act the parts of the Daughters of Sion. The honor of figuring in this holy drama was greatly coveted; and the historian of Dieppe gravely assures us, that the earnestness felt on the occasion mainly contributed to the preservation of that purity of manners and that genuine piety, which subsisted in this town longer than in any other of France! But the election of the Virgin was not sufficient: a representative of St. Peter was also to be found among the clergy; and the laity were so far favored that they were permitted to furnish the eleven other apostles. This done, upon the fourteenth of August the Virgin was laid in a cradle of the form of a tomb, and was carried early in the morning, attended by her suite of either sex, to the church of St. Jacques; while before the door of the master of the guild was stretched a large carpet, embroidered with verses in letters of gold, setting forth his own good qualities, and his love for the holy Mary. Hither also, as soon as Laudes had been sung, the procession repaired from the church, and then they were joined by the governor of the town, the members of the guild, the municipal officers, and the clergy of the parish of St. Remi. Thus attended, they paraded the town, singing hymns, which were accompanied by a full band. The procession was increased by the great body of the inhabitants; and its impressiveness was still farther augmented by numbers of the youth of either sex, who assumed the garb and attributes of their patron saints, and mixed in the immediate train of the principal actors. They then again repaired to the church, where Te Deum was sung by the full choir, in commemoration of the victory over the English, and high mass was performed, and the Sacrament administered to the whole party. During the service, a scenic representation was given of the Assumption of the Virgin. A scaffolding was raised, reaching nearly to the top of the dome, and supporting an azure canopy intended to emulate the "spangled vault of heaven;" and about two feet below the summit of it appeared, seated on a splendid throne, an old man as the image of the Father Almighty, a representation equally absurd and impious, and which could alone be tolerated by the votaries of the worst superstitions of popery. On either side four pasteboard angels of the size of men floated in the air, and flapped their wings in cadence to the sounds of the organ; while above was suspended a large triangle, at whose corners were placed three smaller angels, who, at the intermission of each office, performed upon a set of little bells the hymn of "Ave Maria gratiâ Dei plena per Secula," &c. accompanied by a larger angel on each side with a trumpet. To complete this portion of the spectacle, two others, below the old man's feet, held tapers, which were lighted as the services began, and extinguished at their close; on which occasions the figures were made to express reluctance by turning quickly about; so that it required some dexterity to apply the extinguishers. At the commencement of the mass, two of the angels by the side of the Almighty descended to the foot of the altar, and, placing themselves by the tomb, in which a pasteboard figure of the Virgin had been substituted for her living representative, gently raised it to the feet of the Father. The image, as it mounted, from time to time lifted its head and extended its arms, as if conscious of the approaching beatitude, then, after having received the benediction and been encircled by another angel with a crown of glory, it gradually disappeared behind the clouds. At this instant a buffoon, who all the time had been playing his antics below, burst into an extravagant fit of joy; at one moment clapping his hands most violently, at the next stretching himself out as if dead. Finally, he ran up to the feet of the old man, and hid himself under his legs, so as to shew only his head. The people called him Grimaldi, an appellation that appears to have belonged to him by usage, and it is a singular coincidence that the surname of the noblest family of Genoa the Proud, thus assigned by the rude rabble of a sea-port to their buffoon, should belong of right to the sire and son, whose mops and mowes afford pastime to the upper gallery at Covent-Garden.

Thus did the pageant proceed in all its grotesque glory, and, while—

"These labor'd nothings in so strange a style

Amazed the unlearned, and made the learned smile,"

the children shouted aloud for their favorite Grimaldi; the priests, accompanied with bells, trumpets, and organs, thundered out the mass; the pious were loud in their exclamations of rapture at the devotion of the Virgin; and the whole church was filled with "un non so che di rauco ed indistinto".—But I have told you enough of this foolish story, of which it were well if the folly had been the worst. The sequel was in the same taste and style, and ended with the euthanasia of all similar representations, a hearty dinner.

Footnotes:

[4] *Description de la Haute Normandie, I. p. 130.*

[5] *Histoire de Dieppe, II. p. 86.]*

[6] *Essals sur le Département de la Seine Inférieure, I. p. 119.*

[7] *Histoire de Dieppe, I. p. 1.*

[8] *Another author, mentioned by the Abbé Fontenu, in the Mémoires de l'Académie des Inscriptions, X. p. 413, carries the antiquity of the place still eight centuries higher, representing it as the Portus Ictius, whence Julius Cæsar sailed for Britain.*

[9] *Description de la Haute Normandie, I. p. 125.*

[10] *Vol. XI. p. 55.*

[11] *The deed itself under which this exchange was made is also preserved in Duchesne's Scriptores Normanni, and in the Gallia Christiana, XI. Instr. p. 27, where it is entitled "Celebris commutatio facta inter Richardum I, regem Angliæ et Walterium*

Archiepisc. Rotomagensem." It is worth remarking, in illustration of the feudal rights and customs, how much importance is attached in this instrument to the mills and the seignorage for grinding: the king expressly stipulates that every body *"tam milites quàm clerici, et omnes homines, tam de feodis militum quàm de prebendis, sequentur molendina de Andeli, sicut consueverunt et debent, et moltura erit nostra. Archiepiscopus autem et homines sui de Fraxinis* (a manor specially reserved,) *molent ubi idem Archiepiscopus volet, et si voluerit molere apud Andeli, dabunt molturas suas, sicut alii ibidem molentes. In escambium autem ... concessimus ... omnia molendina quæ nos habuimus Rotomagi, quando hæc permutatio facta fuit, integrè cum omni sequelâ et molturâ suâ, sine aliquo retinemento eorum quæ ad molendinam pertinent vel ad molturam, et cum omnibus libertatibus et liberis consuetudinibus quas solent et debent habere. Nec alicui alii licebit molendinum facere ibidem ad detrimentum prædictorum molendinorum; et debet Archiepiscopus solvere eleemosinas antiquitùs statutas de iisdem molendinis."*

[12]   A very copious and interesting account of the nautical discoveries made by the inhabitants of Dieppe, and of their merits as sailors, is given by Goube, in his *Histoire du Duché de Normandie*, III, p. 172-178.

[13]   Goube, *Histoire de Normandie*, III, p. 170.

[14]   Noel, *Essais sur le Département de la Seine Inférieure*, I. p. 194.

LETTER III
CÆSAR'S CAMP—CASTLE OF ARQUES

(Dieppe, June, 1818)

After having explored Dieppe, I must now conduct you without the walls, to the castle of Arques and to Cæsar's camp, both of which are in its immediate neighborhood. At some future time you may thank me for pointing out these objects to you, for should you ever visit Dieppe, your residence may be prolonged beyond your wishes, by the usual mischances which attend the traveller. And in that case, a walk to these relics of military architecture will furnish a better employment than thumbing the old newspaper of the inn, or even than the contemplation of the diligences as they come in, or of the packets as they are not going out, for I am anticipating that you are becalmed, and that the pennons are flagging from the mast. With respect to my walk, let me be allowed to begin by introducing you to a friend of mine at Dieppe, M. Gaillon, an obliging, sensible, and well-informed young man, as well as an ardent botanist, my companion in this walk, and the source of much of the information I possess respecting these places. The intrenchment, commonly known by the name of Cæsar's camp, or even more generally in the country by that of "la Cité de Limes," and in old writings, of "Civitas Limarum," is situated upon the brink of the cliff, about two miles to the east of Dieppe, on the road leading to Eu, and still preserves in a state of perfection its ancient form and character; though necessarily reduced in the height of its vallum by the operation of time, and probably also diminished in its size by the gradual encroachments of the ocean. Upon its shape, which is an irregular triangle, it may be well to make a preliminary observation, that this was necessarily prescribed by the scite; and that, however the Romans might commonly prefer a square outline for their temporary encampments, we have abundant proofs that they only adhered to this plan when it was perfectly conformable to the nature of the ground, but that when they fortified any commanding position, upon which a rectangular rampart could not be seated, their intrenchments were made to follow the sinuosities of the hill. In the present instance the northern side, the longest, extending nearly five thousand feet, fronts the channel, and it required no other defence than was afforded by the perpendicular face of the cliff, here more than two hundred feet in height. The western side, the second in length, and not greatly inferior to the first, after running about three thousand feet from the sea, in a tolerably straight line southward, suddenly bends to the east, and forms two semi-circles, of one of which the radius is turned from the camp, and of the other into it. The third side is scarcely more than half the length of the others, and runs nearly straight from south to north, where it again unites with the cliff. Of the two last-mentioned sides the first is difficult of access; from its position at the summit of a steep hill; but it is still protected by a vallum from thirty to forty feet high, and between the sea and the entrance nearest to it, a length of about three hundred yards, by a wide exterior ditch with other out-works, as well as by an inner fosse, faint traces of which only now remain. Hence to the next and large entrance is a distance of about two thousand feet; and in this space the interior fosse is still very visible; but the great abruptness of the hill forbade an outer one.

You, who are not a stranger to the pleasures of botany, would have shared my delight at finding upon the perpendicular side of this entrance the beautiful *Caucalis grandiflora*, growing in great luxuriance upon almost bare chalk, and with its snowy flowers resembling, as you look down to it, the common species of *Iberis* of our gardens. The *Asperula cynanchica*, and other plants peculiar to a chalky soil, are also found here in plenty, together with the *Eryngium campestre*, a vegetable of extreme rarity in England, but most abundant throughout the north of France. *Papaver hybridum* is likewise common in the neighboring corn fields round.

Returning from this short botanical digression, let me tell you that the position considered by some as the southern side of the fortification, but which I have described as the sinuous part of the western, has its ramparts of less height. Not so the eastern: on this, as being the most destitute of all natural defence, (for here there is no hill, and the eye ranges over an immense level tract, stopped only by distant woods,) is raised an agger, full forty-five feet in height, and, at a further distance, is added an outward trench nearly fifty feet wide, though in its present state not more than three feet deep, and now serving for a garden.

Such is the external appearance of this camp, which, seen from the sea, or on the approach either by the west or south, cannot fail to strike from the boldness of its position; but the effect of the interior is still more striking; for here, while on one side the horizon is lost in the immensity of the ocean, on the other two the view is narrowly circumscribed by the lofty bulwark, at whose feet are almost every where discernible the remains of the trenches I have already noticed, more than thirty feet in width. Nor is this the only remarkable circumstance; for it is still more unaccountable to observe, extending nearly across the encampment, the traces of an ancient fosse not less than one hundred and fifty feet wide, and, though in most places shallow, terminating towards the sea in a deep ravine. Internally the camp appears to have been also divided into three parts, in one of which it has been supposed, from a heap of stones which till lately remained, that there was originally a place of greater strength; while in another, distinguished by some irregular elevations, it is conjectured that there was a wall, the defence probably to the keep.

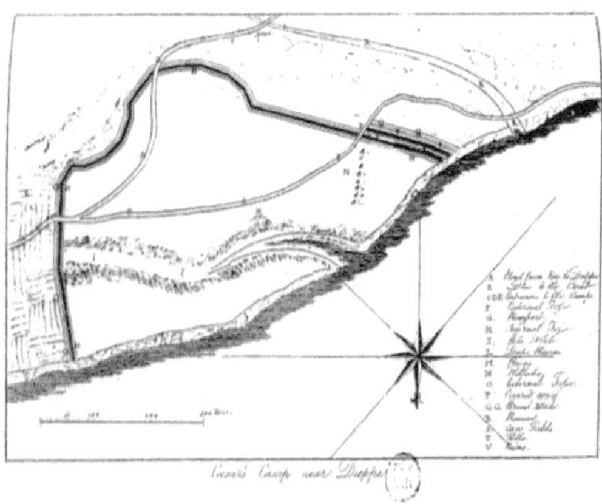
Cæsar's Camp near Dieppe.

But I must tell you that these conjectures are none of my own, nor could I have had any opportunity of making them; the stones and the hillocks having disappeared before the operations of the plough. Such as they are, I have borrowed them from a dissertation by the Abbé de Fontenu[15], a copy of whose engraving of the place I insert. Indebted as I am to him for his hints, I can, however, by no means subscribe to his reasoning, by which he labors with great erudition to prove that, neither the popular tradition which ascribes this camp to Cæsar, nor its name, evidently Roman, nor some coins and medals of the same nation that have been found here, are at all evidences of its Latin origin; but that, as we have no proof that Cæsar was ever in the vicinity of Dieppe, as the whole is in such excellent preservation, (a point I beg leave to deny,) and as the vallum is full thrice the height of that of other Roman encampments in France[16], we are bound to infer it is a work of far more modern times, and probably was erected by Talbot, the Cæsar of the English[17], while besieging Dieppe in the middle of the XVth century.

This opinion of the learned Abbé I quote, principally for the purpose of shewing how far a man of sense and acquirements maybe led astray from truth and probability in support of a favorite theory. Nothing but the love of theory could surely have induced him to suppose that this strong hold was erected for a purpose to which it could in no wise be applicable, as the intervening ground prevents all possibility of seeing any part of Dieppe from the camp, or to ascribe it to times when earth-works were no longer used. In Normandy and Picardy are other camps, more evidently of Roman construction, which are likewise ascribed to Cæsar[18]; with much the same reason perhaps as every thing wonderful in Scotland is referred to Fingal, to King Arthur in Cornwall, and in the north of England and Wales to the devil.

Upon the origin of the castle of Arques, it is somewhat unfortunate for the learned that there is not an equal field for ingenious conjecture, its antiquity being incontestible. Du Moulin, the most comprehensive, though the most credulous of Norman historians, one who, not content with dealing in miracles by wholesale, tells us how the devil changed himself into a postillion, to apprize an alehouse-keeper of the fate of the posterity of Rollo, may still be entitled to credit, when the theme is merely stone and mortar; and from him we may conclude that Arques was a place of importance at the time of William the Conqueror, as it gave the title of Count to his uncle, who then possessed it, and who, confiding perhaps in the strength of his fortress, and secretly instigated by Henry Ist, of France, usurped the title of Duke of Normandy, but was defeated by his nephew, and finally obliged to surrender his castle. This, however, was not till, after a long siege, in which Arques proved itself impregnable to every thing but famine. In the following reign, we again find mention made of Arques, as a portion given by Robert, Duke of Normandy, to induce Helie, son of Lambert of St. Saen, to marry his illegitimate daughter, and join him in defending the Pays de Caux against the English. From this period, during the reigns of the Anglo-Norman Sovereigns, it continues to be occasionally noticed. Before the walls of Arques, according to William of Malmesbury, Baldwin, Count of Flanders, received the wound which afterwards proved fatal. Arques was the last castle which held out in Normandy for King Stephen. It was taken in 1173, by our Henry IInd, and then repaired; was seized by Philip Augustus during the captivity of Richard Cœur de Lion; was restored to its legitimate sovereign at the peace in 1196; and was a source of disgrace to its former captor, when in 1202 he laid siege to it with a powerful army, and was obliged to retreat from its walls. Under the reign of our third Edward, we find it again return to the British crown, as one of the castles specified to be surrendered to the English, by the treaty of Bretigny, in 1359; after which, in 1419, it was taken by Talbot and Warwick, and was finally given up to France by one of the articles of the capitulation of Rouen in 1449. More recently, in 1584[19], it was captured by a party of soldiers disguised like sailors, who, being suffered to approach without distrust, put the sentinels to the sword, and made themselves masters of the fortress; while in 1589 it obtained its last and most honorable distinction, as the chief support of Henry IVth, at the time of his being received at Dieppe, and as having by the cannon from its ramparts, materially contributed to the glorious defeat of the army of the league, commanded by the Duke de Mayenne, when thirty thousand were compelled to retire before one tenth of the number. I have already mentioned to you the address of this king to the citizens of Dieppe: still more magnanimous was his speech to his prisoner, the Count de Belin, previously to this battle, when, on the captive's daring to ask, how with such a handful of men, he could expect to resist so powerful an army, "Ajoutez," he answered, "aux troupes que vous voyez, mon bon droit, et vous ne douterez plus de quel côté sera la victoire."

In Sully's Memoirs[20], as well as in the history of the town of Dieppe, you will find these transactions described at much length, and the warrior, as well as the historian, expatiates on the strength of the castle of Arques; but how much longer it remained a place of consideration I have no means of knowing: most probably the alteration introduced into the art of war by the use of cannon, caused it to be soon after neglected, and dismantled, and suffered to fall gradually into its present state of ruin. It is now the property of a lady residing in the neighboring town of Arques, who purchased it during the revolution, and by her good sense and feeling it has been preserved from further injury. The castle is situated at the extremity of a ridge of chalk hills, which, commencing to the west of Dieppe, run nearly parallel to the sea, and here terminate to the east, so that it has a complete command over the valley. Standing by its walls, you have to the north-west a full view of the town of Dieppe; in an opposite direction the eye ranges uncontrolled over a rich vale of corn and pasturage; and in front, immediately at your feet, lies the town of Arques itself, backed by the hills that are covered by the forest of the same name. Either this forest, or the neighboring one of Eavy, is supposed to have been the ancient Arelanum. The little river called the Arques flows through the valley, and beneath the walls of the castle is lost in the Béthune, under which name the united waters continue their course to Dieppe, after receiving the tribute of a third, yet smaller, stream, the Eaulne.

Of the power of the castle an idea may be formed from the extent of the fosse, little less than half a mile in circumference. The outline of the walls is irregularly oval, and the even front is interrupted by towers of various sizes, and placed at unequal distances. On the northern side, where the hill is steepest, there are no towers; but the walls are still farther strengthened by square buttresses, so large that they indeed look like bastions, and with a projection so great as to indicate an origin posterior to the Norman æra. The two towers which flank the western entrance, and the towers which stand behind each of the flanking towers in the retiring line of the wall, are much larger than any of the rest. One of the latter towers is of so extraordinary a shape, that I consider it as a non-descript; but, as I should tire both you and myself by endeavoring to describe it, I think it most prudent to refer you to a sketch: perhaps its angular parts may not be coeval with the rest of the building[21]: on this it would be impossible to decide positively, so shattered, impaired, and defaced are the walls, and so evidently is their coating the work of different periods. I fancied that in some parts I could discern a mode of construction, in layers of brick and stone, similar to that of Roman buildings in our own country, while many of the bricks, from their texture and shape, appear also to be Roman. Tradition, if we follow that delusive guide, teaches us that we are contemplating a work of the middle of the eighth century, and of one of the sons of Charles Martel. If we follow William of Jumieges, the Chronicle of St. Vandrille, and William of Poitiers, we ascribe it to the uncle and rival of the Conqueror; other writers tell us that the ruins arose under Henry IInd. I dare not decide amongst such reverend authorities, but I think I may infer, without the least disrespect towards monks and chroniclers, that the Norman Arques now occupies the place of a far more early structure, and that a portion of the walls of this latter was actually left in existence. Taken, however, as a whole, the castle is evidently a building of different æras; and it would be extremely difficult, if not impossible, to define the parts belonging to each.

Tower of the Castle of Arques

The principal entrance is to the west, between the two towers first mentioned, over a draw-bridge, whose piers still remain, and through three gateways, whose arches, though now torn and dislocated into shapeless rents, seem to have been circular, and probably of Norman erection. One of the towers of the gate-way appears formerly to have been a chapel. Hence you pass into a court, whose surface, uneven with the remains of foundations, marks it to have been originally filled with apartments, and, at the opposite end of this, through a square gate-house with high embattled walls, a place evidently of great strength, and leading into a large open space that terminated in the quadrangular and lofty keep. This, which is externally strengthened by massy buttresses, similar to those of the walls, is within divided into two apartments, each of them about fifty feet by twenty. In one of them is a well, communicating with a reservoir below, which is filled by the water of the river, and was sufficiently capacious for watering the horses of the garrison. The greatest part, if not the whole, of the walls seems to have been faced with brick of comparatively modern date. The keep also was coated with brick within, and with stones carefully squared without. The windows are so battered, that no idea can be formed of their original style. The walls of the keep are filled with small square apertures. At Rochester, and at many other castles in England, we observe the same; and unless you can give a better guess respecting their use, you must content yourself with mine: that is to say, that they are merely the holes left by the scaffolding. At the foot of the hill to the west is a gate-house, by no means ancient, from which a wall ascends to the castle; and another similar wall connects the fortress with the ground below, on the north-eastern side; but the extent or nature of these out-works can no longer be traced. Still less possible would it be to say any thing with certainty as to the excavations, of the length of which, tradition speaks, as usual, in extravagant terms, and mixes sundry marvellous and frightful tales with the recital.

In the general plan a great resemblance is to be traced between many castles in Wales and its frontiers, especially Goodrich Castle, and this at Arques. Yet I do not think that any of ours are of an equal extent; nor can you well conceive a more noble object than this, when seen at a distance: and it is only then that the eye can comprehend the vast expanse and strength of the external wall, with the noble keep towering high above it.

Church at Arques.

Until the revolution, the decaying town of Arques was not wholly deprived of all the vestiges of its former honours: the standards of the weights and measures of Upper Normandy were deposited here. It was the seat of the courts of the Archbishop of Rouen, and, though the actual session of the municipal courts took place at Dieppe, they bore the legal style and title of the courts of Arques. Since the revolution these traces of its importance have wholly disappeared, nor is there any outward indication of the consequence once enjoyed by this poor and straggling hamlet.

The church is a neat and spacious building, of the same kind of architecture as that of St. Jacques, at Dieppe; and, as it is a good specimen of the florid Norman Gothic, (I forbid all cavils respecting the employment of this term) I have added a figure of it. My slender researches have not enabled me to discover the date of the building, but it may, have been erected towards the year 1350. A most elegant bracket, formed by the graceful dolphin, deserves the attention of the architect; and I particularize it, not merely on account of its beauty, but because, even at the risk of exhausting your antiquarian patience, I intend to point out all architectural features which cannot be retraced in our own structures; and this is one of them. By the way, Arques contributed to increase the bulk of our herbal as well as of our sketch-book, for under the walls of the church is found the rare *Erodium moschatum*; and near the castle grow *Astragalus glycyphyllos* and *Melissa Nepeta*.

The field of battle is to the southward of the town. A small walk under the south wall of the castle, near the east end, adjoining a covered way which led to a postern-gate or draw-bridge, is still called the walk of Henry the IVth, because it was here that this monarch was wont to reconnoitre the enemy's forces from below.

Napoléon, towards the conclusion of his reign, visited the field of battle at Arques; he ascertained the position of the two armies, and pronounced that the King ought to have lost the day, for that his tactics were altogether faulty. I am willing to suppose that this military criticism arose merely from military pedantry, though it is now said that Napoléon was envious of the veneration, which, as the French believe, they feel for the memory of Henri quatre. Napoléon is accused of having given the title of *le Roi de la Canaille* to the Bourbon Monarch. And when Napoléon was in full-blown pride, he might have had the satisfaction of hearing the rabble of Paris chaunt his comparative excellence in a parody of the old national song—

> "Vive Bonaparte, vice ce conquérant,
>
> Ce diable à quatre a bien plus de talent
>
> Que ce Henri quatre et tous ses descendans,"

Footnotes:

[15] Mémoires de l'Académie des Inscriptions, X. p. 403. tab. 15.

[16] Such are the Abbé's principal arguments; but he goes on to say, that the height of the ramparts proves almost to demonstration their having been erected since the use of fire-arms, a mode of reasoning that would, I fear, be equally conclusive against the antiquity of a very celebrated earth-work, the Devil's-Ditch, in Cambridgeshire, whose agger is of about the same elevation, but of whose modern origin nobody ever yet dreamed;—that the ramparts opposite

Dieppe could only be of use against cannon, another position equally untenable;—that, were the camp Roman, there would be platforms on the agger for the reception of wooden towers, as if time would not wear away vestiges of this nature;—that the disposition is not in regular order like that of a Roman encampment, a matter equally liable to be defaced;—and, finally, that the out-works to the west are fully decisive of a more modern æra, as if intrenchments were not, like buildings, frequently the objects of subsequent alterations;—In his inferences he is followed, and, apparently without any question as to their authenticity, by Ducarel, whom I suspect from his description never to have visited the place. The Abbé Fontenu, in a paper in the same volume, gives it as his opinion that, from the term Civitas Limarum, it might safely be believed there was a city in this place; and he tries to persuade himself that he can trace the foundations of houses.

[17] Noel, Essais sur le Départment de la Seine Inférieure, I. p. 88.

[18] The same is also notoriously the case in our own country: popular tradition, by a metonymy very easily to be accounted for, from a desire of adding importance to its objects, attributes whatever is Roman to Julius Cæsar, as the most illustrious of the Roman generals in England; just as we daily hear smatterers in art referring to Raphael any painting, however ordinary, that pretends to issue from the schools of Rome or Florence, every Bolognese one to Guido or Annibal Carracci, every Kermes to Ostade or Teniers, &c.

[19] Noel, Essais sur la Seine Inférieure, I. p. 98.

[20] Sully, who was himself in this battle, and bore a conspicuous part in it, dwells upon its details completely con amore, and evidently regards the issue of this day as decisive of the fate of the monarch, who is reported to have said of himself shortly before the battle, that "he was a king without a kingdom, a husband without a wife, and a warrior without money."—I. p. 204.

[21] In justice to my readers, I must not here omit to say that such is the opinion of a most able friend of mine, Mr. Cohen, who visited this castle nearly at the same time with myself, and who writes me on the subject: "I feel convinced

that the brick coating of the wedge-tower at Arques is recent. Such was the impression I had upon the spot; and now I cannot remove it. It appeared to me that the character of the brick-work, and of the stone cordons or fillets, was entirely like that of the fortifications of the XVIth century; and I also thought, perhaps erroneously, that the wedge or bastion was affixed to the round tower of the castle, and that it was an after-construction. At the south end of the castle, you certainly see very ancient and singular masonry. The diagonal or herring-bone courses are found in the old church of St. Lo, and in the keep at Falaise; not in the front of the latter, but on the side where you enter, and on the side which ranges with Talbot's Tower. The same style of masonry is also seen, according to Sir Henry Englefield, at Silchester, which is most undoubtedly a pure Roman relic." —It abounds likewise in Colchester Castle.

LETTER IV
JOURNEY FROM DIEPPE TO ROUEN—PRIORY OF LONGUEVILLE—ROUEN—BRIDGE OF BOATS—COSTUME OF THE INHABITANTS

(Rouen, June, 1818.)

I arrived alone at this city: my companions, who do not always care to keep pace with my constitutional impatience, which sometimes amuses, and now and then annoys them, made a circuit by Havre, Bolbec, and Yvetot, while I proceeded by the straight and beaten track. What I have thus gained in expedition, I have lost in interest. During the whole of the ride, there was not a single object to excite curiosity, nor would any moderate deviation from the line of road have brought me within reach of any town or tower worthy of notice, except the Priory of Longueville, situate to the right of the road, about twelve miles from Dieppe. I did not see Longueville, and I am told that the ruins are quite insignificant, yet I regret that I did not visit them. The French can never be made to believe that an old rubble wall is really and truly worth a day's journey: hence their reports respecting the notability of any given ruin can seldom be depended upon. And at least I should have had the satisfaction of ascertaining the actual state of the remains of a building, known to have been founded and partly built in the year 1084, by Walter Giffard[22], one of the relations and companions of the Conqueror, in his descent upon England, and therefore created Earl of Buckingham, or, as the French sometimes write it, Bou Kin Kan. The title was held by his family only till 1164 when, upon the decease of his son without issue, the lands of his barony were shared among the collateral female heirs. He himself died in 1102, and by his will directed that his body should be brought here, which was accordingly done; and he was buried, as Ordericus Vitalis[23] tells us, near the entrance of the church, having over him an epitaph of eight lines, "in maceriâ picturis decoratâ." You will find the epitaph, wherein he is styled "templi fundator et ædificator," copied both in the Neustria Pia and in Ducarel's Anglo-Norman Antiquities. The latter speaks of it as if it existed in his time; but the doctor seldom states the extent of his obligations towards his predecessors. And in consequence of this his silent gratitude, we can never tell with any degree of certainty whether we are perusing his observations or his transcripts. If he really saw the inscriptions with his own eyes, it is greatly to be regretted that he has given us no information respecting the paintings: did they still exist, they would afford a most genuine and curious proof of the state of Norman art at that remote period; and possibly, a search after them among the cottages in the neighborhood might even now repay the industry of some keen antiquary; for the French revolution may well he compared to an earthquake: it swallowed up every thing, ingulphing some so deep that they are lost for ever, but leaving others, like hidden treasures, buried near the surface of the soil, whence accident and labor are daily bringing them to light. The descendants of Walter Giffard are repeatedly mentioned as persons of importance in the early Norman writers; nor are they less illustrious in England, where the great family of Clare sprung from one of the daughters; while another, by her marriage with Richard Granville, gave birth to the various noble families of that name, of which the present Marquis of Buckingham is the chief.

Of the Priory, we are told in the Neustria Pia, that it was anciently of much opulence, and that a Queen of France contributed largely to the endowment of the house. Many men of eminence, particularly three of the Talbot family, were buried within its walls. Peter Megissier, a prior of Longueville, was in the number of the judges who passed sentence of death upon the unfortunate Joan of Arc; and the inscription upon his tomb is so good a specimen of monkish Latinity, that I am tempted to send it you; reminding you at the same time, that this barbarous system of rhyming in Latin, however brought to perfection by the monks and therefore generally called their own, is not really of their invention, but may be found, though quoted to be ridiculed, in the first satire of Persius,

> "Qui videt hunc lapidem, cognoscat quòd tegit idem
>
> Petrum, qui pridem conventum rexit ibidem
>
> Annis bis senis, tumidis Leo, largus egenis,
>
> Omnibus indigenis charus fuit atque alienis."

I believe it is always expected, that a traveller in France should say something respecting the general aspect of the country and its agriculture. I shall content myself with remarking, that this part of Normandy is marvellously like the country which the Conqueror conquered. When the weather is dull, the Normans have a sober English sky, abounding in Indian ink and neutral tint. And when the weather is fine, they have a sun which is not a ray brighter than an English sun. The hedges and ditches wear a familiar livery, and the land which is fully cultivated repays the toil of the husbandman with some of the most luxuriant crops of wheat I ever saw.

Barley and oats are not equally good, perhaps from the stiffness of the soil, which is principally of chalk; but flax is abundant and luxuriant. The surface of the ground is undulated, and sufficiently so to make a pleasing alternation of hill and dale; hence it is agreeably varied, though the hills never rise to such a height as to be an obstacle to agriculture. There is some difficulty in conjecturing where the people by whom the whole is kept in cultivation are housed; for the number of houses by the road-side is inconsiderable; nor did we, for the first two-thirds of the ride, pass through a single village, excepting Tôtes, which lies mid-way between Dieppe, and Rouen, and is of no great extent. Yet things in France are materially altered in this respect since 1814, when I remember that, in going through Calais by the way of the Low Countries to Paris, and returning by the direct road to Boullogne, the whole journey was made without seeing a single new house erecting in a space of four hundred miles. This is now far from being the case; there is every where an appearance of comparative prosperity, and, were it not for the coins, of which the copper bear the impress of the republic, and the gold and silver chiefly that of Napoléon, a stranger would meet with but few visible marks of the changes experienced in late years by the government of France. Much has been also done of late towards ornamenting the châteaux, of which there are several about Tôtes, though in the opinion of an Englishman, much also is yet wanting. They are principally the residences of Rouen merchants.

Upon approaching Malaunay, about nine miles from Rouen, the scene is entirely changed. The road descends into a valley, inclosed between steep hills, whose sides are richly and beautifully clothed with wood, while the houses and church of the village beneath add life and variety to the plain at the foot. Here the cotton manufactories begin, and, as we follow the course of the little river Cailly, the population gradually increases, and continues to become more dense through a series of manufacturing villages, each larger than the preceding, and all abounding in noble views of hill, wood, and dale; while the tracts around are thickly studded with picturesque residences of manufacturers, and extensive, often picturesque, manufactories. Such indeed was the country, till we found ourselves at Rouen, shortly before entering which the Havre road unites to that from Dieppe, and the landscape also embraces the valley of the Seine, as well as of the Cailly the former broader by far, and grander, but not more beautiful.

Rouen, from this point of view, is seen to considerable advantage, at least by those who, like us, make a *détour* to the north, and enter it in that direction: the cathedral, St. Ouen, the hospital and church of La Madeleine, and the river, fill the picture; nor is the impression in any wise diminished on a nearer approach, when, through a long avenue, formed by four rows of lofty elms, you advance by the side of a stream, at once majestic from its width and eminently beautiful from its winding course.

Rouen is now unfortified; its walls, its castles, are level with the ground. But, if I may borrow the pun of which old Peter Heylin is guilty when, describing Paris, Rouen is still a strong city, "for it taketh you by the nose." The filth is extreme; villainous smells overcome you in every quarter, and from every quarter. The streets are gloomy, narrow, and crooked, and the houses at once mean and lofty. Even on the quay, where all the activity of commerce is visible, and where the outward signs of opulence might be expected, there is nothing to fulfil the expectation. Here is width and space, but no trottoir; and the buildings are as incongruous as can well be imagined, whether as to height, color, projection, or material. Most of them, and indeed most in the city, are merely of lath and plaster, the timbers uncovered and painted red or black, the plaster frequently coated with small grey slates laid one over another, like the weather-tiles in Sussex. Their general form is very tall and very narrow, which adds to the singularity of their appearance; but mixed with these are others of white brick or stone, and really handsome, or, it might be said, elegant. The contrast, however, which they form only makes their neighbors look the more shabby, while they themselves derive from the association an air of meanness. The merchants usually meet upon a small open plot, situated opposite to the quay, inclosed with palisades and fronted with trees. This is their exchange in fine weather; but adjoining is a handsome building, called La Bourse à couvert, or Le Consulte, to which recourse is always had in case of rain. It was here that Napoléon and Maria Louisa, a very short time previous to their deposition, received from the inhabitants of Rouen the oath of allegiance, which so soon afterwards found a ready transfer to another sovereign.

About the middle of the quay is placed the bridge of boats, an object of attraction to all strangers, but more so from the novelty and singularity of its construction than from its beauty. Utility rather than elegance was consulted by the builder. This far-famed structure is ugly and cumbrous, and a passenger feels a very unpleasing sensation if he happens to stand upon it when a loaded waggon drives along it at low water, at which time there is a considerable descent from the side of the suburbs. An undulatory motion is then occasioned, which goes on gradually from boat to boat till it reaches the opposite shore. The bridge is supported upon nineteen large barges, which rise and fall with the tide, and are so put together that one or more can easily be removed as often as it is necessary to allow any vessel to pass. The whole too can be entirely taken away in six hours, a construction highly useful in a river peculiarly liable to floods from sudden thaws; which sometimes occasion such an increase of the waters, as to render the lower stories of the houses in the adjacent parts of the city uninhabitable. The bridge itself was destroyed by a similar accident, in 1709, for want of a timely removal. Its plan is commonly attributed to a monk of the order of St. Augustine, by whom it was erected in 1626, about sixty years after the stone bridge, built by the Empress Matilda in 1167, had ceased to be passable. It seems the fate of Rouen to have wonderful bridges. The present is dignified by some writers with the high title of a miracle of art: the former is said by Taillepied, in whose time it was standing, to have been "un des plus beaux édifices et des plus admirables de la France." A few lines afterwards, however, this ingenuous writer confesses that loaded carriages of any kind were seldom suffered to pass this admirable edifice, in consequence of the expence of repairing it; but that two barges were continually plying for the transport of heavy goods. The delay between the destruction of the stone bridge, and the erection of the boat bridge, appears to have been occasioned by the desire of the citizens to have a second similar to the first; but this, after repeated deliberations, was at last determined to be impracticable, from the depth and rapidity of the stream. Napoléon, however, seems to have thought that the task which had been accomplished under the auspices of the Empress Matilda, might be again repeated in the name of the daughter of the Cæsars and the wife of the successor of Charlemagne; and he actually caused Maria-Louisa to lay the first stone of a new bridge, at some distance farther to the east, where an island divides the river into two. This, I am told, will certainly he finished, though at an enormous expence, and though it will occasion great inconvenience to many inhabitants of the quay, whose houses will be rendered useless by the height to which it will be necessary to raise the soil upon the occasion. My informant added, that, small as is the appearance yet made above water, whole quarries of stone and forests of wood have been already sunk for the purpose.

From the scite of the projected bridge, the view eastward is particularly charming. The bold hill of St. Catherine presents its steep side of bare chalk, spotted only in a few places with vegetation or cottages, and seems to oppose an impassable barrier; the mixture of country-houses with trees at its base, makes a most pleasing variety; and, still nearer, the noble elms of the boulevards add a character of magnificence possessed by few other cities. The boulevards of Rouen are rather deficient in the Parisian accompaniments of dancing-dogs and music-grinders, but the sober pedestrian will, perhaps, prefer them to their namesakes in the capital. Here they are not, as at Paris, in the centre of the town, but they surround it, except upon the quay, with which they unite at each end, and unite most pleasingly; so that, immediately on leaving this brilliant bustling scene, you enter into the gloom of a lofty embowered arcade, resembling in appearance, as well as in effect, the public walks at Cambridge, except that the addition of females in the fanciful Norman costume, and of the Seine, and the fine prospect beyond, and Mont St. Catherine above, give it a new interest. On the opposite side of the Seine, the inhabitants of Rouen have another excellent promenade in the grand cours, which, for a considerable space, occupies the bank of the river, turning eastward from the bridge. Four rows of trees divide it into three separate walks, of which the central one is by far the widest, and serves for horses and carriages; the other two are appropriated exclusively to foot passengers. In these, on a summer's evening, are to be seen all classes of the inhabitants of Rouen, from the highest to the lowest; and the following sketch, which you will easily perceive to be from a pencil more delicate than mine, gives a most lively and faithful picture of them. It may indeed be in some measure in the nature of a treatise de re vestiariá, yet such details of gowns and petticoats never fail to interest, at least to interest me, when proceeding from a wearer.

"Our carriage had scarcely stopped when we were surrounded with

beggars, principally women with children in their arms. The poor babes presented a most pitiable appearance, meagre, dirty to the utmost degree, ragged and flea-bitten, so that round the throat there was not the least portion of "carnation" appearing to be free from the insect plague. Their hair, too, is seldom cut; and I have seen girls of eight or ten years of age, bearing a growing crop which had evidently remained unshorn, and I may add, uncombed, from the time of their birth. It is impossible not to dread coming into contact with these imps, who, when old, are among the ugliest conceivable specimens of the human race. The women, even those who inhabit the towns, live much in the open air: besides being employed in many slavish offices, they sit at their doors or windows pursuing their business, or lounge about, watching passengers to obtain charity. Thus their faces and necks are always of a copper color, and, at an advanced age, more dusky still; so that, for the anatomy and coloring of witches, a painter needs look no further. Their wretchedness is strongly contrasted by the gaiety of the higher classes. The military, who, I suppose, as usual in France, hold the first place, appear in all possible variety of keeping and costume, with their well-proportioned figures, clean apparel, decided gait, martial air, and whiskered faces. Here and there we see gliding along the well-dressed lady (not well dressed, indeed, as far as becomingness goes, but fashionably), with a gown of triple flounces, whose skirt intrudes even upon the shoulders, obliterating the waist entirely, while her throat is lost in an immense frill of four or more ranks; and sometimes a large shawl over all completes the disguise of the shape. The head of the dame or damsel is usually enveloped in a gauze or silk bonnet, sufficiently large to spread, were it laid upon a table, two feet in diameter, and trimmed with various-colored ribbons and artificial flowers: in the hand is seen the ridicule, a never-failing accompaniment. The lower orders of women at Rouen usually wear the Cauchoise cap, or an approach to it, rising high to a narrowish point at top, and furnished with immense ears or wings that drop on the shoulder, then opening in front so as to allow to be seen on the forehead a small portion of hair, which divides and falls in two or three spiral ringlets on each side of the face. The remainder of the dress is generally composed of a colored petticoat, probably striped, an apron of a different color, a bodice still differing in tint from the rest, and a shawl, uniting all the various hues of all the other parts of the dress. Some of the peasants from the country look still more picturesque, when mounted on horseback bringing vegetables: they keep their situation without saddle or stirrup, and seem perfectly at ease. But the best figures on horseback are the young men who take out their masters' horses to give them exercise, and who are frequently seen on the grand cours. They ride without hat, coat, saddle, or saddle-cloth, and with the shirt sleeves rolled up above the elbow. Their negligent equipment, added to their short, curling hair, and the ease and elasticity they display in the management of their horses, gives them, on the whole, a great resemblance to the Grecian warriors of the Elgin marbles. Men, as well as women, are frequently seen without hats in the streets, and continually uncravatted; and when their heads are covered, these coverings are of every shape and hue; from the black beaver, with or without a rim, through all gradations of cap, to the simple white cotton nightcap. A painter would delight in this display of forms and these sparkling touches of color, especially when contrasted with the grey of the city, and the tender tints of the sky, water, and distance, and the broad coloring of the landscape."

Footnotes:

[22] *"He was son of Osborne de Bolebec and Aveline his wife, sister to Gunnora, Duchess of Normandy, great-grandmother to the Conqueror, and was one of the principal persons who composed the general survey of the realm, especially for the county of Worcester. In 1089 he adhered to William Rufus, against his brother Robert Courthose, and forfeited his Norman possessions on the king's behalf, of whose army there he was a principal commander, and behaved himself very honorably. Yet, in the time of Henry Ist, he took the part of the said Courthose against that king, but died the year following,"*—Banks' *Extinct Baronagé,* III. p. 108.

[23] *Duchesne, Scriptores Normanni, p. 809.*

LETTER V
JOURNEY TO HAVRE—PAYS DE CAUX—ST. VALLERY—FÉCAMP—THE PRECIOUS BLOOD—THE ABBEY—TOMBS IN IT—MONTIVILLIERS—HARFLEUR

(Rouen, June, 1818.)

Lest I should deserve to be visited with the censure which I have taken the liberty of passing upon Ducarel's tour, I shall begin by premising that my account of the present state of the tract, intended for the subject of this and the following letter, is wholly derived from the journals of my companions. Their road by Fécamp, Havre, Bolbec, and Yvetot, has led them through the greater part of the Pays de Caux, a district which, in the time of Cæsar, was peopled by the Caletes or Caleti. Antiquaries suppose, that in the name of this tribe, they discover the traces of its Celtic origin, and that its radical is no other than the word Kalt or Celt itself. As a proof of the correctness of this etymology, Bourgueville[25] tells us that but little more than two hundred years have passed since its inhabitants, now universally called Cauchois, were not less commonly called Caillots or Caillettes; a name which still remains attached to several families, as well as to the village Gonfreville la Caillotte, and, probably, to some others. I shall, however, waive all Celtic theory, "for that way madness lies," and enter upon more sober chorography.

The author of the Description of Upper Normandy states, that the territory known by that appellation was limited to the Pays de Caux and the Vexin: the former occupying the line of sea-coast from the Brêle to the Seine, together with the governments of Eu and Havre and the Pays de Brai; the latter comprising the Roumois, and the French as well as the Norman Vexin. All these territorial divisions have, indeed, been obliterated by the state-geographers of the revolution; and Normandy, time-honored Normandy herself, has disappeared from the map of the dominions of the French king. The ancient duchy is severed into the five departments of the Seine Inférieure, the Eure, the Orne, Calvados, and the Manche. These are the only denominations known to the government or to the law, yet they are scarcely received in common parlance. The people still speak of Normandy, and they still take a pleasure in considering themselves as Normans: and, I too, can share in their attachment to a name, which transmits the remembrance of actual sovereignty and departed glory.

Until the re-union of feudal Normandy to the crown of its liege lord, the duke was one of the twelve peers of the kingdom; and to his hands that kingdom entrusted the sacred Oriflamme, as often as it was expedient to unfurl it in war. Normandy also contained several titular duchies, ancient fiefs held of the King as Duke of Normandy, but which, out of favour to their owners, were "erected," as the French lawyers say, into duchies, after the province had reverted to the crown. This erection, however, gave but a title to the noble owner, without increasing his territorial privileges; nor could any of our Richards, or our Henries, have allowed a liege man to write himself duke, like his proud feudal suzerein. The recent duchies were Alençon, Aumale, Harcourt, Damville, Elbeuf, Etouteville, and Longueville, and three of them were included in the Pays de Gaux, the inhabitants of which, from the titles connected with it, were accustomed to dignify it with the epithet of noble. Their claim to the epithet is thus given by an ancient Norman poet of the fifteenth century; and if, according to the old tradition, which Voltaire has bantered with his usually incredulity, we could admit that Yvetot was ever really a kingdom, it must be allowed that few provinces could produce such a titled terrier:

> "Au noble Pays de Caux
>
> Y a quatre Abbayes royaux,
>
> Six Prieurés conventionaux,
>
> Et six Barons de grand arroi,
>
> Quatre Comtes, trois Ducs, un Roi."

The soil of the district is generally rich; but the farmers frequently suffer from drought, especially in its western part, where they are obliged almost constantly to have recourse to artifical irrigation. The houses and villages are all surrounded with hedges, thickly planted, and each village is also belted in the same manner. These inclosures, which are peculiar to the Pays de Caux, give a monotonous appearance to the landscape, but they are highly beneficial, for they break the force of the winds, and furnish the inhabitants with fuel. If my memory does not deceive me, the towns either of the ancient Gauls or Teutons, are described as being thus encompassed in primitive times; but I cannot name my authorities for the assertion.

St. Vallery, the first stage beyond Dieppe, is situated in a valley; and there is an obscure tradition that this valley was once watered by a river, which disappeared some centuries ago. It is conjectured, from the name of the town, that it claims an origin as high as the seventh century, when the disciples of St. Vallery were obliged to quit their original monastery and take refuge elsewhere. Yet, according to other authorities[26], it did not receive its present appellation till 1197, when Richard Cœur de Lion, after having destroyed the town and abbey of St. Vallery sur Somme, carried off the relics of the patron saint, and deposited them in this town. My reporters tell me that it has an air of antiquity and gloom, but that it contains nothing worthy of notice except a crucifix in the churchyard, of stone, richly wrought, dated 1575, and a bénitier of such simple form and rude workmanship, as to appear of considerable antiquity. The place itself is only a wretched residence for four or five thousand fishermen; but still it has a name[27] in history. Hence William sailed for the conquest of England; and its harbor, all poor and small as it is, has always been considered of importance to the country; there being no other between Havre and Dieppe capable of affording shelter to vessels of even a moderate size.

The road to Fécamp passes through the little town of Cany, situated in a beautiful valley; and there my family met the Archbishop of Rouen, who, at this moment, is in progress through his diocese, for the purpose of confirmation. The approach of his eminence gave the appearance of a fair to every village: young and old of both sexes were collected in the highways to welcome the prelate. He travelled in considerable state, attended by a military escort of twenty men; and arrayed in the scarlet robe of a Roman Cardinal, with the brilliant "decoration" of the Legion of Honor conspicuous upon his breast. For the archbishop is a grand officer of that brotherhood of bastard chivalry; and this ornament, conjoined to his train of whiskered warriors, seemed to render him a very type of the church militant. His eminence is extremely bulky; and my pilgrims were wicked enough to be much amused by the oddity of his pomp and pride. Nor did the postillion spare his facetiousness on the occasion; for you are aware that in France, as in most other parts of the continent, the servile classes use a degree of familiarity in their intercourse with their betters, to which we are little accustomed in England, and which has given rise to the Italian proverb, that "Il Francese è fedele, l'Italiano rispettoso, l'Inglese schiavo[28]."

Throughout this part of France, large flocks of sheep are commonly seen in the vicinity of the sea, and, as the pastures are uninclosed, they are all regularly guarded by a shepherd and his black dog, whose activity cannot fail to be a subject of admiration. He is always on the alert and attentive to his business, skirting his flock to keep them from straggling, and that, apparently, without any directions from his master. In the night they are folded upon the ploughed land; and the shepherd lodges, like a Tartar in his kibitka, in a small cart roofed and fitted up with doors.

Fécamp, like other towns in the neighborhood, is imbedded in a deep valley; and the road, on approaching it, threads through an opening between hills "stern and wild," a tract of "brown heath and shaggy wood," resembling many parts of Scotland. The town is long and straggling, the streets steep and crooked; its inhabitants, according to the official account of the population of France, amount to seven thousand, and the number of its houses is estimated at thirteen hundred, besides above a third of that quantity which are deserted, and more or less in ruins[29].

Fécamp appeared desolate and decaying to its visitors, but they recollected that its very desolation was a voucher of the antiquity from which it derives its interest. It claims an origin as high as the days of Cæsar, when it was called *Fisci Campus*, being the station where the tribute was collected.

It is in vain, however, to expect concord amongst etymologists; and, of course, there are other right learned wights who protest against this derivation. They shake their heads and say, "no; you must trace the name, Fécamp, to Fici Campus;" and they strengthen their assertion by a sort of argumentum ad ecclesiam, maintaining that the precious blood, for which Fécamp was long celebrated, corroborates and confirms their tale. A chapel in the abbey church attests the sanctity of this relic. The legend states that Nicodemus, at the time of the entombment of our Saviour, collected in a phial the blood from his wounds, and bequeathed it to his nephew, Isaac; who afterwards, making a tour through Gaul, stopped in the Pays de Caux, and buried the phial at the root of a fig-tree[30].

Nor is this the only miracle connected with the church. The monkish historians descant with florid eloquence upon the white stag, which pointed out to Duke Ansegirus the spot where the edifice was to be erected; the mystic knife, inscribed "in nomine sanctæ et individuæ trinitatis," thus declaring to whom the building should be dedicated; and the roof, which, though prepared for a distant edifice, felt that it would be best at Fécamp, and actually, of its own accord, undertook a voyage by sea, and landed, without the displacing of a single nail, upon the sea-coast near the town. All these contes dévots, and many others, you will find recorded in the Neustria Pia. I will only detain you with a few words more upon the subject of the precious blood, a matter too important to be thus hastily dismissed. It was placed here by Duke Richard I.; but was lost in the course of a long and turbulent period, and was not found again till the year 1171, when it was discovered within the substance of a column built in the wall. Two little tubes of lead originally contained the treasure; but these were soon inclosed in two others of a more precious metal, and the whole was laid at the bottom of a box of gilt silver, placed in a beautiful pyramidical shrine. Thus protected, it was, before the revolution, fastened to one of the pillars of the choir, behind a trellis-work of copper, and was an object of general adoration. I know not what has since become of it; but, as they are now managing these matters better in France, we may safely calculate upon the speedy reappearance of the relic. Nor must you refer this legend to the many which protestant incredulity is too apt to class with the idle tales of all ages, the

> "... quicquid Græcia mendax
>
> Audet in historiâ;"

for no less grave an authority than the faculty of theology at Paris determined, by a formal decree of the 28th of May, 1448, that this worship was very proper; for that, to use their words, "Non repugnat pietati fidelium credere quòd aliquid de sanguine Christi effuso tempore passionis remanserit in terris."

The abbey, to which Fécamp was indebted for all its greatness and celebrity, was founded in 664[32] for a community of nuns, by Waning, the count or governor of the Pays de Caux, a nobleman who had already contributed to the endowment of the Monastery of St. Wandrille. St. Ouen, Bishop of Rouen, dedicated the church in the presence of King Clotaire; and, so rapidly did the fame of the sanctity of the abbey extend, that the number of its inmates amounted in a very short period to three hundred or more. The arrival, however, of the Normans, under Hastings, in 841, caused the dispersion of the nuns; and the same story is related of the few who remained at Fécamp, as of many others under similar circumstances, that they voluntarily cut off their noses and their lips, rather than be an object of attraction to the lust of their conquerors. The abbey, in return for their heroism, was levelled with the ground, and it did not rise from its ashes till the year 988, when the piety of Duke Richard I. built the church anew, under the auspices of his son, Robert, Archbishop of Rouen; but, departing from the original foundation, he established therein a chapter of regular canons, who, however, were so irregular in their conduct, that within ten years they were doomed to give way to a body of Benedictine Monks, headed by an Abbot, named William, from a convent at Dijon. From his time the monastery continued to increase in splendor. Three suffragan abbies, that of Notre Dame at Bernay, of St. Taurin at Evreux, and of Ste. Berthe de Blangi, in the diocese of Boullogne, owned the superior power of the abbot of Fécamp, and supplied the three mitres which he proudly bore on his abbatial shield.

Kings and princes in former ages frequently paid the abbey the homage of their worship and their gifts; and, in a period nearer to our own, Casimir of Poland, after his voluntary abdication of the throne, selected it as the spot in which he sought for repose, when wearied with the cares of royalty. The English possessions of Fécamp (for like most of the great Norman abbeys, it held lands in our island) do not appear to have been large; but, according to an author of our own country[33] the abbot presented to one hundred and thirty benefices, some in the diocese of Rouen, others in those of Bayeux, Lisieux, Coutances, Chartres, and Beauvais; and it enjoyed so many estates, that its income was said to be forty thousand crowns per annum. Fécamp moreover could boast of a noble library, well stored with manuscripts[34], and containing among its archives many original charters, deeds, &c. of William the Conqueror, and several of his successors.

This magnificent church is three hundred and seventy feet long and seventy high; the transept, including the Chapel of the Precious Blood, one hundred and twenty feet long; the tower two hundred feet high. A portion of it was burned in 1460, but soon repaired. William de Ros, third abbot, rebuilt all the upper part in a better taste, and enlarged the nave, which was not finished till 1200. A successor of his at the beginning of the next century completed the chapels round the choir. The screen was begun by one of the monks about 1500, who erected the chapel dedicated to the death of the Virgin, a master-piece of architecture and adorned with historical carving. The cloister was built so late as 1712. Cathedral service was performed in the church, in which were the tombs of the first and second of the Richards of Normandy; of Richard, infant son of the former, and of William, third son of the latter; of Margaret, betrothed to Robert, son of William the Conqueror, who died 1060; of Alard, third Earl of Bretagne, 1040; of Archbishop Osmond, and of a Lady Judith, whose jingling epitaph has given rise to a variety of conjectures, whether she was the wife of Duke Richard IInd, or his daughter, or some other person. —

> "Illa solo sociata, mariti at jure soluta,
>
> Judita judicio justificata jacet;
>
> Et quæ, dante Deo, sed judice justificante,
>
> Primo jus subiit sed modò jura regit."

As to Duke Richard Ist, he caused a sarcophagus of stone to be made and placed within this church; and so long as he lived, it was filled with wheat on every Friday, and the grain, together with five shillings, distributed weekly among the poor. And when his death approached, he expressly charged his successor, "Bury not my body within the church, but deposit it on the outside, immediately under the eaves, that the dripping of the rain from the holy roof may wash my bones as I lie, and may cleanse them of the spots of impurity contracted during a negligent and neglected life."

Our party could not ascertain whether any of the historical monuments were yet in existence. The church, at the time they were there, was wholly occupied with preparations for the approaching confirmation. Young girls in their best dresses, all in white, and holding tapers in their hands, filled the nave, while the chapels were crowded with individuals at prayer, or still more with females waiting for an opportunity of confessing themselves, previously to receiving the expected absolution from the archbishop. Under such circumstances nothing could be examined; but there appeared to be in the chapels five or six fine, though mutilated, altar tombs: to whom, however, they belonged, or what was their actual state, it was impossible to tell. Accompanying them are also some curious pieces of sculpture. For the same reason no farther remark could be made upon the interior of the building, except that its architecture is imposing, and its roof, supported by tall clustered pillars, has much the general effect of the nave of our cathedral at Norwich, one of the purest specimens of Norman architecture in England. Externally the tower is handsome, and of nearly the earliest pointed style; not altogether so, as its arches, though narrow, contain each a double arch within. The rest of the building seems to have suffered much from alterations and dilapidation; and whatever tracery there may have been originally has disappeared from the windows; nor are there saints or even niches remaining above the doors.

The exterior of the church of St. Etienne, one of the ten parochial churches of Fécamp, before the revolution, is considerably more imposing; but upon this I will not detain you, as you will see it engraved in Mr. Cotman's *Architectural Antiquities of Normandy,* from a sketch taken by him last year.

Henry IInd, of England, made a donation of the town to the abbey, whose seignorial jurisdiction also extended over many other parishes, as well in this as in the adjoining dioceses. Its exclusive privileges were likewise ample. Under the first and second race, Fécamp was the seat of government of the Pays de Caux, and the residence of the counts of the district: it was also a residence of the Norman Dukes. Their castle was rebuilt by William Longue-Epeé, with a degree of magnificence which is said to have been extraordinary. This duke took particular pleasure in the place, and he and his immediate successors frequently lived here. But the palace has long since disappeared[35]: the continual increase of the monastic buildings gradually occupied its place; and they, in their turn, are now experiencing the revolutions of fortune, the inhabitants being at this very time actively employed in their demolition.

The town is at present wholly supported by the fisheries, in which are employed about fourteen hundred sailors[36]. The herrings of Fécamp have always had the same high character in France, as those of Lowestoft and Yarmouth in England. The armorial lion of our own town ends, as you know, with the tail of a herring; and I really have been often inclined to affix the same appendage to the rump of the lion of Normandy. You are not much of an epicure, nor are you very likely to search in the Almanach des Gourmands for dainties; if you did, you would probably find there the following proverb, which has existed since the thirteenth century,—

    "Aloses de Bourdeaux;

    Esturgeons de Blaye;

    Congres de la Rochelle;

    Harengs de Fécamp;

    Saumons de Loire;

    Sêches de Coutances."

The fortifications of Fécamp are destroyed; but, upon the cliffs which command the town, there still remain some slight vestiges of a fort, erected in the time of Henry IVth, when the inhabitants espoused the party of the league. The capture of this fort was one of those gallant exploits which the historian delights in recording; and it is detailed at great length in Sully's Memoirs[37].

From Fécamp to Havre the country is well wooded, and much applied to the cultivation of flax, which flourishes in this neighborhood, and has given rise to considerable linen manufactories. The trees look well in masses, but individually they are trimmed into ugliness. Near Havre the road goes through Montivilliers, and, still nearer, through Harfleur.

The first of these is, like Fécamp, a place of antiquity, and derived its name[38] and importance from a monastery which was founded at the end of the seventh century. Its history is headed by the chapter which begins the records of most of the ecclesiastical foundations of the duchy: when the invading heathen Normans reached Montivilliers, it shared the common fate of destruction, and when they withdrew, the common piety recalled it to existence. Richard IInd bestowed it upon Fécamp, but the same sovereign restored it to its independence, at the request of his aunt, Beatrice, who retired hither as abbess, at the head of a community of nuns. A convent, over which an abbess of royal blood had presided, could not fail to enjoy considerable privileges; and it retained them to the period of the revolution. The tower of the church still remains, a noble specimen of the Norman architecture of the eleventh century, at which period the building is known to have been erected. The rest of the edifice, though handsome as a whole, is the work of different æras. The archives of the monastery furnish an account of large sums expended in additions and alterations in the years 1370 and 1513. The interior contains some elegant stone fillagree-work in the form of a small gallery or pulpit, attached to the west end near the roof, and probably intended to receive a band of singers on high festivals. A gallery of a similar nature, but of wood, and to which the foregoing purpose was assigned by the learned wight, John Carter, is yet remaining at the north-west corner of Westminster Abbey. You and I, who are sadly inclined to admire ugliness and antiquity, would have been better pleased with the capitals of the pillars, which are evidently coeval with the tower. Drawings were made of some of these capitals, and I have selected two which appeared to be the most singular.

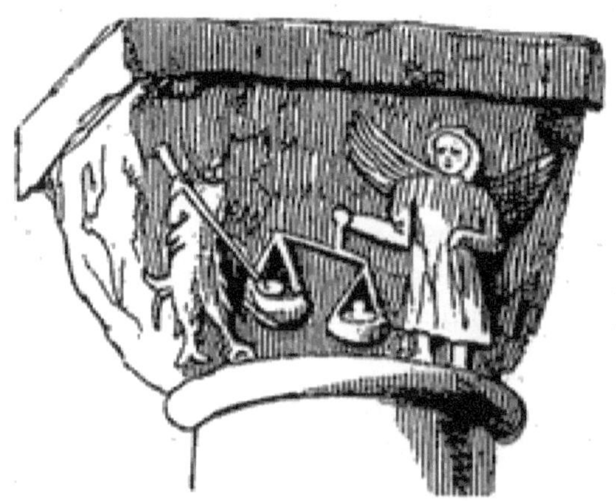

In this you observe an angel weighing the good works of the deceased against his evil deeds; and, as the former are far exceeding the avoirdupois upon which Satan is to found his claim, he is endeavoring most unfairly to depress the scale with his two-pronged fork.

This allegory is of frequent occurrence in the monkish legends.—The saint, who was aware of the frauds of the fiend, resolved to hold the balance himself.—He began by throwing in a pilgrimage to a miraculous virgin.—The devil pulled out an assignation with some fair mortal Madonna, who had ceased to be immaculate.—The saint laid in the scale the sackcloth and ashes of the penitent of Lenten-time.—Satan answered the deposit by the vizard and leafy-robe of the masker of the carnival.—Thus did they still continue equally interchanging the sorrows of godliness with the sweets of sin, and still the saint was distressed beyond compare, by observing that the scale of the wicked thing (wise men call him the correcting principle,) always seemed the heaviest. Almost did he despair of his client's salvation, when he luckily saw eight little jetty black claws just hooking and clenching over the rim of the golden basin. The claws at once betrayed the craft of the cloven foot. Old Nick had put a little cunning young devil under the balance, who, following the dictates of his senior, kept clinging to the scale, and swaying it down with all his might and main. The saint sent the imp to his proper place in a moment, and instantly the burthen of transgression was seen to kick the beam.

Painters and sculptors also often introduced this ancient allegory of the balance of good and evil, in their representations of the last judgment: it was even employed by Lucas Kranach.

The other capital which I send to you is ornamented with groups of Centaurs or Sagittaries. Astronomical sculptures are frequently found upon the monuments of the middle ages. Two capitals, forming part of a series of zodiacal sculptures, are preserved in the *Musée des Monumens Français;* and, speaking from memory, I think they bear a near resemblance in style to that which is here represented.

Montivilliers itself is a neat little town, beautifully situated in a valley, with a stream of clear water running through it. At this time its trade is trifling; but the case was otherwise in former days, when its cloths were considered to rival those of Flanders, and the preservation of the manufacture was regarded of so much consequence, that sundry regulations respecting it are to be found in the royal ordinances. One of them in particular, of the fourteenth century, notices the frauds committed by other towns in imitating the mark of the cloth of Montivilliers.

The general appearance of Harfleur is much like that of Montivilliers; but numerous remains of walls and gates denote that it was once of still greater comparative importance. The ancient trade of the place is now transferred to Havre de Grace, the situation of the latter town being far more elegible.

The Seine no longer rolls its waves under Harfleur; and the desiccated harbor is now seen as a verdant meadow. Without the aid of history, therefore, you would in vain inquire into the derivation of the name, in connection with which, the learned Huet, Bishop of Avranches[39], calls upon us to remark, that the names of many places in Normandy end in fleur, as Barfleur, Harfleur, Honfleur, Fiefleur, Vitefleur, &c.; and that, if, as it is commonly supposed, this termination comes from fluctus, it must have passed through the Saxon, in which language fleoten signifies to flow. Hence we have flot, and from flot, fleut and fleur, the last alteration being warranted by the genius of the French language. The bishop further states, that there are two facts, affording a decisive proof of this origin: the one, that the names now terminating in fleur, ended anciently flot, Barfleur being Barbeflot, Harfleur Hareflot, and Honfleur Huneflot; the other, that all places so called are situated where they are washed by the tide. Such is also the position of the towns in Holland, whose names terminate in vliet, and of those in England, ending in fleet, as Purfleet, Byfleet, &c. The Latin word flevus is of the same kind, and is derived from the same source; for, instead of Hareflot and Huneflot, some old records have Hareflou and Huneflou, and some others Barfleu, terms approaching flevus, which is also called by Ptolemy, fleus, and by Mela, fletio. It is highly improbable, that these two last terms should have been coined subsequently to the time of the Romans becoming masters of Gaul, and it is equally unlikely that the Saxon fleoten should be derived from the Latin. Thus far, therefore, the languages appear to have had a common origin, and they are insomuch allied to the Celtic, that those towns in Britanny, in whose names are found the syllables pleu and plou, are also invariably placed in similar situations.

If, however, I am fairly embarked in the sea of etymological conjecture, I know not where I shall be carried; and therefore, instead of urging the probability that the root of the Celtic pleu is apparently to be found in the Pelasgic πλεω, I shall return to Harfleur and its history. Whilst Harfleur was in its glory, it was considered the key of the Seine and of this part of France. In 1415 it opposed a vigorous resistance to our Henry Vth, who had no sooner made himself master of it, than, with a degree of contradiction, which teaches man to regard the performance of his duty to God as no reason for his performing it to his fellow-creatures, "the King uncovered his feet and legs, and walked barefoot from the gate to the parish church of St. Martin, where he very devoutly offered up his prayers and thanksgivings for his success. But, immediately afterwards he made all the nobles and the men at arms that were in the town his captives, and shortly after sent the greater part out of the place, clothed in their jerkins only, taking down their names and surnames in writing, and obliging them to swear by their faith that they would surrender themselves prisoners at Calais on Martinmas-day next ensuing. In like manner were the townsmen made prisoners, and obliged to ransom themselves for large sums of money. Afterwards did the King banish them out of the town, with numbers of women and children, to each of whom were given five sols and a portion of their garments." Monstrelet[40], from whom I have transcribed this detail, adds, that "it was pitiful to hear and see the sorrow of these poor people, thus driven away from their homes; the priests and clergy were likewise dismissed; and, in regard to the wealth found there, it was not to be told, and appertained even to the King, who distributed it as he pleased." Other writers tell us that the number of those thus expelled was eight thousand, and that the conqueror, not satisfied with this act of vengeance, publicly burned the charters and archives of the town and the title-deeds of individuals, re-peopled Harfleur with English, and forbad the few inhabitants that remained to possess or inherit any landed property. After a lapse, however, of twenty years, the peasants of the neighboring country, aided by one hundred and four of the inhabitants, retook the place by assault. The exploit was gallant; and a custom continued to prevail in Harfleur, for above two centuries subsequently, intended to commemorate it; a bell was tolled one hundred and four times every morning at day-break, being the time when the attack was made. In 1440, the citizens, undismayed by the sufferings of their predecessors, withstood a second siege from our countrymen, whom the town resisted four months, and in whose possession it remained ten years, when Charles VIIIth permanently united it to the crown of France. Notwithstanding these calamities, it rose again to a state of prosperity, till the revocation of the edict of Nantes gave the death-blow to its commerce; and intolerance completed the desolation which war had begun. At present, it is only remarkable for the elegant tower and spire of its church, connected by flying buttresses of great beauty, the whole of rich and elaborate workmanship.

At a short distance from Harfleur, the Seine comes in view, flowing into the sea through a fine rich valley; but the wide expanse of water has no picturesque beauty. The hills around Havre are plentifully spotted with gentlemen's houses, few only of which have been seen in other parts in the ride. The town itself is strongly fortified; and, having conducted you hither, I shall leave you for the present, reserving for another letter any particulars respecting Havre, and the rest of the road to Rouen.

Footnotes:

[25] *Antiquités de Normandie, p. 53.*

[26] *Dumoulin, Géographie de la France, II p. 80.*

[27] *Description de la Haute Normandie, I. p. 109.*

[28] *Heylin notices the familiarity of the approach of the French servants, in his delineation of a Norman inn. An extract may amuse those who are not familiar with the works of this quaint yet sensible writer. "There stood in the chamber three beds, if at the least it be lawful so to call them; the foundation of them was straw, so infinitely thronged together, that the wool-packs which our judges sit on in the Parliament, were melted butter to them; upon this lay a medley of flocks and feathers sewed up together in a large bag, (for I am confident it was not a tick) but so ill ordered that the knobs stuck out on each side like a crab-tree cudgel. He had need to have flesh enough that lyeth on one of them, otherwise the second night would wear out his bones. —Let us now walk into the kitchen and observe their provision. And here we found a most terrible execution committed on the person of a pullet; my hostess, cruel woman, had cut the throat of it, and without plucking off the feathers, tore it into pieces with her hands, and afterwards took away skin and feathers together: this done, it was clapped into a pan and fried for supper. —But the principal ornaments of these inns are the men-servants, the raggedest regiment that ever I yet looked upon; such a thing as a chamberlain was never heard of amongst them, and good clothes are as little known as he. By the habits of his attendants a man would think himself in a gaol, their clothes are either full of patches or open to the skin. Bid one of them make clean your boots, and presently he hath recourse to the curtains. —They wait always with their hats on, and so do all servants attending on their masters. —Time and use reconciled me to many other things, which, at the first were offensive; to this most irreverent custom I returned an enemy; neither can I see how it can choose but stomach the most patient to see the worthiest*

*sign of liberty usurped and profaned by the basest of slaves."—
Peter then has a learned excursus de jure pileorum, wherein
Tertullian de Spectaculis, Erasmus his Chiliades, and many other
reverent authorities are adduced; also, giving an account of his
successful exertions, as to "the licence of putting on our caps at
our public meetings, which privilege, time, and the tyranny of the
vice-chancellor, had taken from." After which, he still resumes in
ire,—"this French sauciness hath drawn me out of the way; an
impudent familiarity, which, I confess, did much offend me; and
to which I still profess myself an open enemy. Though Jacke speak
French, I cannot endure Jacke should be a gentleman."*

[29] *Géographie de la France, II. p. 115.*

[30] *Description de la Haute Normandie, I. p. 94.*

[32] *Description de la Haute Normandie, I. p. 90.—Some other
writers date the foundation A.D. 666.*

[33] *Gough's Alien Priories, I. p. 9.*

[34] *This important part of its treasures, we may hope, from
the following passage in Noel, has been in a measure preserved.
"On m'a assuré que cette dernière partie des richesses littéraires
de notre pays étoit heureusement conserveé: puisse aujourd'hui
ce dépot, honorant les mains qui le possédent, parvenir intégre
jusqu'aux tems propères où le génie de l'histoire pourra utiliser sa
possession."—Essais sur la Seine Inférieure, II. p. 21.*

[35] *I do not know if it be wholly destroyed; for the author of
the Description of Upper Normandy and Goube both speak of the
existence of a square tower within the precincts of the abbey, part
of the old palace, and known by the name of the Tower of Babel.*

[36] *Noel, Essais sur la Seine Inférieure, II. p. 11.*

[37] *Vol. I. p. 389.*

[38] *This name, in Latin, is Monasterium Villare; in old French
records it is called Monstier Vieil.*

[39] *Origines de Caen, 2nd edit. p. 300.*

[40] *Vol. II. p. 78.*

LETTER VI
HAVRE—TRADE AND HISTORY OF THE TOWN—EMINENT MEN—BOLBEC—YVETOT—RIDE TO ROUEN—FRENCH BEGGARS

(Rouen, June, 1818.)

To Fécamp and the other places noticed in my last letter, a more striking contrast could not easily be found than Havre. It equally wants the interest derived from ancient history, and the appearance of misery inseparable from present decay. And yet even Havre is now suffering and depressed. A town which depends altogether upon foreign commerce, could not fail to feel the effects of a long maritime war; and we accordingly find the number of its inhabitants, which twenty years ago was estimated at twenty-five thousand, now reduced to little more than sixteen thousand.

The blow, which Havre will with most difficulty recover is the loss of St. Domingo; for, before the revolution, it almost enjoyed a monopoly of the trade of this important colony, in which upwards of eighty ships, each of above three hundred tons burthen, were constantly employed. With Martinique and Guadaloupe it had a similar, though less extensive, intercourse. As the natural outlet for the manufactures of Rouen and Paris, it supplied the French islands in the West Indies with the principal part of their plantation stores; and the situation of the port was equally advantageous for the importation of their produce. Guinea and the coast of Africa afforded a second and important branch of commerce; and this also is little likely entirely to recover. We may add that, happily it is not so; for it depended principally upon the slave-trade, the profits of which were such, that it was calculated a vessel might clear upon an average nearly eight thousand pounds by each voyage[41]. Its whale-fishery has, for more than a century, ceased to exist. This pursuit began with spirit and at as early a period as the year 1632, when the merchants of this port, in conjunction with those of Biscay, fitted out the expedition commanded by Vrolicq, seized upon a station near Spitzbergen, where they would have obtained a permanent establishment, had they not been violently expelled by the

Danes and Dutch. But the coasting-trade with the various ports of France, and the communication with the other countries of Europe, is now again in full vigor; and it is to these sources that Havre is chiefly indebted for the life and spirit visible in its quays and public places.

The appearance of bustle and activity is a striking, at the same time that it is a most pleasing, character, of every great and commercial sea-port, in every part of the world: it is especially so in a climate which is milder than our own, and where not only the loading and unloading of the ships, with the consequent transport of merchandize, is continually taking place before the spectator; but the sides of the shops are commonly set open, sail-makers are pursuing their business in rows in the streets, and almost every handicraft and occupation is carried on in the open air. An acute traveller might also conjecture that the mildness of the atmosphere is comfortable and congenial to the parrots, perroquets, and monkeys, which are brought over as pets and companions by the sailors. Great numbers of these exotic birds and brutes are to be seen at the windows, and they almost give to the town of Havre the appearance of a tropical settlement.

The quays are strongly edged and faced with granite: the streets, of which there are forty, are all built in straight lines, and chiefly at right angles with each other. In them are several fountains, round which picturesque groups of women are continually collected, employed with Homeric industry in the task of washing linen. The churches are ugly, their style is a miserable caricature of Roman architecture, the interiors are incumbered by dirty and dark chapels, filled up with wood carvings. The principal church has figures of saints, of wretched execution, but of the size of life, ranged round the interior. The harbor is calculated to contain three hundred vessels. The houses are oddly constructed: they are very narrow, and very lofty, being commonly seven stories high, and they are mostly fronted with stripes of tiled slate, and intermediate ones of mortar, so fantastically disposed, that two are rarely seen alike.

Notwithstanding what is alledged by the author of the Mémoires sur Havre, in his endeavors to give consequence to his native place, by maintaining its antiquity, it appears certain that no mention is made of the town previously to the fifteenth century. Even so late as 1509, its scite was occupied by a few hovels, clustered round a thatched chapel, under the protection of Notre Dame de Grace, from whom the place derived the name of Havre de Grace. Francis Ist, who was the real founder[42] of Havre, was desirous of changing this name to Françoisville or Franciscopole. But the will of a sovereign, as Goube very justly observes, most commonly dies with him: in our days, the National Convention, aided by the full force of popular enthusiasm, has equally failed in a similar attempt. The jacobins tried in vain to banish the recollections of good St. Denis, by unchristening his vill under the appellation of Franciade. Disobedience to the edict, exposed, indeed, the contravener to the chance of experiencing the martyrdom of the bishop; yet the mandate still produced no effect. Nor was Napoléon more successful; and history affords abundant proof, that it is more easy to build a city, or even to conquer a kingdom, than to alter an established name.

Viewed in its present condition, no town in France unites more advantages than Havre: it is one of the keys of the kingdom; it commands the mouth of the river that leads direct to the metropolis; and it is at once a great commercial town and a naval station. Possessing such claims to commercial and military pre-eminence, it may appear matter of surprise that it should be of so recent an origin; but the cause is to be sought for in the changes which succeeding centuries have induced in the face of the country —

"Vidi ego quæ fuerat quondam durissima tellus

Esse fretum; vidi factas ex æquore terras."

The sea continually loses here, and, without great efforts on the part of man to retard the operation of the elements, Havre may, in process of time, become what Harfleur is. At its origin it stood immediately on the shore; the consequence of which was, that, within a very few years, a high tide buried two-thirds of the houses and nearly all the inhabitants. The remembrance of this dreadful calamity is still annually renewed by a solemn procession on the fifteenth of January.

With regard to historical events connected with Havre, there is little to be said. It was the spot whence our Henry VIIth embarked, in 1485, aided by four thousand men from Charles VIIIth, of France, to enforce his claim to the English crown. The town was seized by the Huguenots, and delivered to our Queen Elizabeth, in 1562. But it was held by her only till the following year, when Charles IXth, with Catherine of Medicis, commanded the siege in person, and pressed it so vigorously, that the Earl of Warwick was obliged to evacuate the place, after having sacrificed the greater part of his troops. At the end of the following century, after the bombardment and destruction of Dieppe, an attack was made upon Havre, but without success, owing to the strength of the fortifications, and particularly of the citadel. For this, the town was indebted to Cardinal Richelieu, who was its governor for a considerable time, and who also erected some of its public buildings, improved the basin, and gave a fresh impulse to trade, by ordering several large ships of war to be built here. As ship-builders, the inhabitants of Havre have always had a high character: they stand conspicuous in the annals of the art, for the construction of the vessel called la Grande Françoise, and justly termed la grande, as having been of two thousand tons burthen. Her cables are said to have been above the thickness of a man's leg; and, besides what is usually found in a ship, she contained a wind-mill and a tennis-court[43]. Her destination was, according to some authors, the East Indies; according to others, the Isle of Rhodes, then attacked by Soliman IInd; but we need not now inquire whither she was bound; for, after advantage had been taken of two of the highest tides, the utmost which could be done was to tow her to the end of the pier, where she stuck fast, and was finally obliged to be cut to pieces. Her history and catastrophe are immortalized by Rabelais, under the appellation of la Grande Nau Françoise.

It were unpardonable to take leave of Havre without one word upon the celebrated individuals to whom it has given birth; and you must allow me also, from our common taste for natural history, to point it out to your notice as a spot peculiarly favorable for the collecting of fossil shells, which are found about the town and neighborhood in great numbers and variety. The Abbé Dicquemare, a naturalist of considerable eminence, who resided here, may possibly be known to you by his observations on this subject, or still more probably by those upon the Aetiniæ; the latter having been translated into English, and honored with a place in the Transactions of our Royal Society. Of more extensive, but not more justly merited, fame, are George Scudery and his sister Magdalen: the one a voluminous writer in his day, though now little known, except for his Critical Observations upon the Cid; the other, a still more prolific author of novels, and alternately styled by her contemporaries the Sappho of her age, and "un boutique de verbiage;" but unquestionably a writer of merit, notwithstanding the many unmanly sneers of Boileau, whose bitter pen, like that of our own illustrious satirist, could not even consent to spare a female that had been so unfortunate as to provoke his resentment. She died in 1701, at the advanced age of ninety-four. The last upon my list is one of whom death has very recently deprived the world, the excellent Bernardin de Saint Pierre; a man whose writings are not less calculated to improve the heart than to enlarge the mind. It is impossible to read his works without feeling love and respect for the author. His exquisite little tale of Paul and Virginia is in the hands of every body; and his larger work, the Studies of Nature, deserves to be no less generally read, as full of the most original observations, joined to theories always ingenious, though occasionally fanciful: the whole conveyed in a singularly captivating style, and its merits still farther enhanced by a constant flow of unaffected piety.

The road from Havre to Rouen is of a different character, and altogether unlike that from Dieppe; but what it gains in beauty of landscape it loses in interest. And yet, perhaps, it is even wrong to say that it gains much in point of beauty; for, though: trees are more generally dispersed, though cultivation is universal, and the soil good, and produce luxuriant, and though the mind and the eye cannot but be pleased by the abundance and verdure of the country, yet in picturesque effect it is extremely deficient. Monotony, even of excellence, displeases. I am speaking of the road which passes through Bolbec and Yvetot: there is another which lies nearer to the banks of the Seine, through Lillebonne and Caudebec, and this, I do not doubt, would, in every point of view, have been preferable.

At but a short distance from Havre, to the left, lies the church, formerly part of the priory, of Grâville, a picturesque and interesting object. Of the date of its erection we have no certain knowledge, and it is much to be regretted that we have not, for it is clearly of Norman architecture; the tower a very pure specimen of that style, and the end of the north transept one of the most curious any where to be seen, and apparently; also one of the most ancient[44]. I should therefore feel no scruple in referring the building to a more early period than the beginning of the thirteenth century, where our records of the establishment commence; for it was then that William Malet, Lord of Grâville, placed here a number of regular canons from Ste. Barbe en Auge, and endowed them with all the tythes and patronage he possessed in France and England. The act by which Walter, Archbishop of Rouen, confirmed this foundation, is dated in 1203. Stachys Germanica, a plant of extreme rarity in England, grows abundantly here by the road-side; and apple-trees are very numerous, not only edging the road, but planted in rows across the fields.

The valley by which you enter Bolbec is pretty and varied; full of trees and houses, which stand at different heights upon the hills on either side. The town itself is long, straggling, and uneven. Through it runs a rapid little stream, which serves many purposes of extensive business, connected with the cotton manufactory, the preparation of leather, cutlery, &c. This stream, of the same name with the town, afterwards falls into the Seine, near Lillebonne, one of the most ancient places in Normandy, and formerly the metropolis of the Caletes, but now only a wretched village. Tradition refers its ruin to the period of the invasion of Gaul by the Romans; but it revived under the Norman Dukes, who resided here a portion of the year, and it was a favorite seat of William the Conqueror. To him, or to one of his immediate predecessors or successors, it is most probable that the castle owes its existence. Mr. Cotman found the ruins of it extensive and remarkable. The importance of the place, at a far more early date, is proved by the medals of the Upper and Lower Empire, which are frequently dug up here, and not less decisively by the many Roman roads which originate from the town.

Bolbec can lay claim to no similar distinction; but it is full of industrious manufacturers. Twice in the last century it was burned to the ground; and, after each conflagration, it has arisen more flourishing from its ashes. At the last, which happened in 1765, Louis XVth made a donation to the town of eighty thousand livres, and the parliament of Normandy added a gratuity of half as much more, to assist the inhabitants in repairing their losses.

Yvetot, the next stage, possesses no visible interest, and furnishes no employment for the pencil. The town is, like Bolbec, a residence for manufacturers; and the curious stranger would seek in vain for any traces of decayed magnificence, any vestiges or records of a royal residence. And yet, it is held that Yvetot was the capital of a *kingdom*, which, if it really did exist, had certainly the distinction of being the smallest that ever was ruled on its own account. The subject has much exercised the talents and ingenuity of historians. It has been maintained by the affirmants, that an actual monarchy existed here at a period as remote as the sixth century; others argue that, though the Lords of Yvetot may have been stiled *Kings*, the distinction was merely titular, and was not conferred till about the year 1400; whilst a third, and, perhaps, most numerous, body, treat the whole as apocryphal.

Robert Gaguin[45], a French historian of the fifteenth century, prefaces the anecdote by observing, that he is the first French writer by whom it is recorded; and, as if sensible that such a remark could not fail to excite suspicion, he proceeds to say, that it is wonderful that his predecessors should have been silent. Yet he certainly was not the first who stated the story in print; for it appears in the Chronicles of Nicholas Gilles, which were printed in 1492, whilst the earliest edition of Gaugin was published in 1497.—According to these monkish historians, Clotharius, of France, son of Clovis, had threatened the life of his chamberlain, Gaultier, Lord of Yvetot, who thereupon fled the kingdom, and for ten years remained in voluntary exile, fighting against the infidels. At the end of this period, Gaultier hoped that the anger of his sovereign might be appeased, and he accordingly went to Rome, and implored the aid of the Supreme Pontiff. Pope Agapetus pitied the wanderer; and he gave unto him a letter addressed to the King of the Franks, in which he interceded for the supplicant. Clotharius was then residing at Soissons, his capital, and thither Gaultier repaired on Good-Friday, in the year 536, and, availing himself of the moment when the King was kneeling before the altar, threw himself at the feet of the royal votary, beseeching pardon in the name of the common Savior of mankind, who on that day shed his blood for the redemption of the human race. But his prayers and appeal were in vain: he found no pardon; Clothair drew his sword, and slew him on the spot. The Pope threatened the monarch with apostolical vengeance, and Clothair attempted to atone for the murder, by raising the town and territory of Yvetot into a kingdom, and granting it in perpetuity to the heirs of Gaultier.

Such is the tradition. There is a very able dissertation upon the subject, by the Abbé de Vertot[46], who endeavors to disprove the whole story: first by the silence of all contemporary authors; then by the fact, that Yvetot was not at that time under the dominion of Clothair; then by an anachronism, which the story involves as to Pope Agapetus; and finally by sundry other arguments of minor importance. Even he, however, admits, that in a royal decree, dated 1392, and preserved among the records of the Exchequer of Normandy, the title of King is given to the Lord of Yvetot; and he is obliged to cut the knot, which he is unable to untie, by stating it as his opinion, that at or about this period Yvetot was really raised into a sovereignty, though, on what occasion, for what purpose, and with what privileges, no document remains to prove. As a parallel case, he instances the Peers of France, an order with whose existence every body is acquainted, while of the date of the establishment nothing is known. It is surprising, that so clear-sighted a writer did not perceive that he was doing nothing more than illustrating, as the logicians say, obscurum per obscurius, or, rather, making darkness more dark; as if it were not considerably more probable, that so strange a circumstance should have taken place in the sixth century, and have been left unrecorded, when society was unformed, anomalies frequent, and historians few, than that it should have happened in the fourteenth, a period when the government of France was completely settled in a regular form, under one monarch, when literature was generally diffused, and when every remarkable event was chronicled. Besides which, the inhabitants of the little kingdom continued, in some measure, independent of his Most Christian Majesty, even until the revolution. At least, they paid not a sou of taxes, neither aides, nor tenth-penny, nor gabelle. It was a sanctuary into which no farmer of the revenue dared to enter. And it is hardly to be doubted, but that there must have been some very singular cause for so singular and enviable a privilege. In our own days, M. Duputel[47], a member of the academy of Rouen, has entered the lists against the Abbé; and between them the matter is still undecided, and is likely so to continue. For myself, I have no means of throwing light upon it; but the impression left upon my mind, after reading both sides of the question, is, that the arguments are altogether in favor of Vertot, while the greater weight of probabilities is in the opposite scale. I shall leave you, however, to poise the balance, and I shall not attempt to cause either end of the beam to preponderate, by acting the part of Old Nick as before exhibited to you; though I decidedly believe that Gaguin had some authority for his tale, but, by neglecting to quote it, he has left the minds of his readers to uncertainty, and his own veracity to suspicion.

With this digression I bid farewell to Yvetot, and its Lilliputian kingdom; nor will I detain you much longer on the way to Rouen, the road passing through nothing likely to afford interest in point of historical recollection or antiquities; though within a very short distance of the ancient Abbey of Pavilly on the one side, and at no great distance from the still more celebrated Monastery of Jumieges on the other. The houses in this neighborhood are in general composed of a framework of wood, with the interstices filled with clay, in which are imbedded small pieces of glass, disposed in rows, for windows. The wooden studs are preserved from the weather by slates, laid one over the other, like the scales of a fish, along their whole surface, or occasionally by wood over wood in the same manner. I am told that there are some very ancient timber churches in Norway, erected immediately after the conversion of the Northmen, which are covered with wood-scales: the coincidence is probably accidental, yet it is not altogether unworthy of notice. At one end the roof projects beyond the gable four or five feet, in order to protect a door-way and ladder or staircase that leads to it; and this elevation has a very picturesque effect. A series of villages, composed of cottages of this description, mixed with large manufactories and extensive bleaching grounds, comprise all that is to be remarked in the remainder of the ride; a journey that would be as interesting to a traveller in quest of statistical information, as it would be the contrary to you or to me.

Poverty, the inseparable companion of a manufacturing population, shews itself in the number of beggars that infest this road as well as that from Calais to Paris. They station themselves by the side of every hill, as regularly as the mendicants of Rome were wont to do upon the bridges. Sometimes a small nosegay thrown into your carriage announces the petition in language, which, though mute, is more likely to prove efficacious than the loudest prayer. Most commonly, however, there is no lack of words; and, after a plaintive voice has repeatedly assailed you with "une petite charité, s'il vous plait, Messieurs et Dames," an appeal is generally made to your devotion, by their gabbling over the Lord's Prayer and the Creed with the greatest possible velocity. At the conclusion, I have often been told that they have repeated them once, and will do so a second time if I desire it! Should all this prove ineffectual, you will not fail to hear "allons, Messieurs et Dames, pour l'amour de Dieu, qu'il vous donné un bon voyage," or probably a song or two; the whole interlarded with scraps of prayers, and ave-marias, and promises to secure you "santé et salut." They go through it with an earnestness and pertinacity almost inconceivable, whatever rebuffs they may receive. Their good temper, too, is undisturbed, and their face is generally as piteous as their language and tone; though every now and then a laugh will out, and probably at the very moment when they are telling you they are "pauvres petits misérables," or "petits malheureux, qui n'ont ni père ni mère." With all this they are excellent flatterers. An Englishman is sure to be "milord," and a lady to be "ma belle duchesse," or "ma belle princesse." They will try too to please you by "vivent les Anglais, vive Louis dix-huit." In 1814 and 1815, I remember the cry used commonly to be "vive Napoléon," but they have now learned better; and, in truth, they had no reason to bear attachment to the ex-emperor, an early maxim of whose policy it was to rid the face of the country of this description of persons, for which purpose he established workhouses, or dépots de mendicité, in each department, and his gendarmes were directed to proceed in the most summary manner, by conveying every mendicant and vagrant to these receptacles, without listening to any excuse, or granting any delay. He had no clear idea of the necessity of the gentle formalities of a summons, and a pass under his worship's hand and seal. And, without entering into the elaborate researches respecting the original habitat of a mumper, which are required by the English law, he thought that pauperism could be sufficiently protected by consigning the specimen to the nearest cabinet. The simple and rigorous plan of Napoléon was conformable to the nature of his government, and it effectually answered the purpose. The day, therefore, of his exile to Elba was a Beggar's Opera throughout France; and they have kept up the jubilee to the present hour, and seem likely to persist in maintaining it.

Footnotes:

[41] *Goube, Histoire de la Normandie, III. p. 127.*

[42] *"François premier, revenant vainqueur de la bataille de Marignan en 1515, crut devoir profiter de la situation avantageuse de la Crique; il conçut le dessin de l'agrandir et d'en faire une place de guerre importante. Ce prince avoit pris les interêts du jeune Roi d'Ecosse, Jacques V, et ce fut pour se fortifier contre les Anglais qu'il forma la résolution de leur opposer cette barrière. Pour conduire l'entreprise il jetta les yeux sur un Gentilhomme nommé Guion le Roi, Seigneur de Chillon, Vice-Amiral, et Capitaine de Honfleur, et la premiere pierre fut posée en 1516."—Description de la Haute Normandie, I. p. 195.*

[43] *Description de la Haute Normandie, I. p. 200.*

[44] *See Cotman's Architectural Antiquities of Normandy, t. 12.—There is also a general view of the church, and of some of the monastic buildings from the lithographic press of the Comte de Lasteyrie.*

[45] *"Sed priusquàm a Clotario discedo, illud non prætermittendum reor, quod, cùm maximè cognitu dignum est, mirari licet a nullo Franco Scriptore litteris fuisse commendatum. Fuit inter familiarissimos Clotarii aulicos, Galterus Yvetotus, Caletus agri Rothomagensis, apprimè nobilis et qui regii cubiculi primarius cultor esset. Huic pro suâ integritate, de Clotario cùm meliùs meliùsque in dies promereretur, reliqui aulici invident, depravantes quodlibet ab eo gestum, nec desistunt donec irritatum illi Clotarium pessimis susurris efficiunt; quamobrem jurat Rex se hominem necaturum. Perceptâ Clotarii indignatione, Galterus pugnator illustris cedere Regi irato constituit. Igitur derelictâ Franciâ in militiam adversus religionis catholicæ inimicos pergit, ubi decem annos multis prosperè gestis rebus, ratus Clotarium simul cum tempore mitiorem effectum, Romam in primis ad Agapitum Pontificem se contulit: a quo ad Clotarium impetratis litteris, ad eum Suessione agentem se protinùs confert, Veneris die, quæ parasceve dicitur, cogitans religiosam Christianis diem ad pietatem sibi profuturam. Verùm litteris Pontificis exceptis cùm Galterum Clotarius agnovit, vetere irâ tanquam recenti livore percitus, rapto a proximo sibi equite gladio, hominem statìm interemit. Tam indignam insignis atque innocentis hominis necem, religioso loco et die ad Christi passionem recolendam celebri, pontifex inæquanimitèr ferens, confestìm Clotarium reprehendit, monetque iniquissimi facinoris rationem habere, se alioquin excommunicationis sententiam subiturum. Agapiti monita reveritus Rex, capto cum prudentibus consilio, Galteri hæredes, et qui Yvetotum deinceps possiderent, ab omni Francorum Regum ditione atque fide liberavit, liberosque prorsùs fore suo syngrapho et regiis scriptis confirmat. Ex quo factum est ut ejus pagi et terræ possessor Regem se Yvetoti hactenus sine controversiâ nominaverit. Id autem anno christianæ gratiæ quingentesimo trigesimo sexto gestum esse indubiâ fide invenio. Nam dominantibus longo post tempore in Normanniâ. Anglis, ortâque inter Joannem Hollandum, Auglum, et Yvetoti dominum quæstione, quasi proventuum ejus terræ pars fisco Regis Anglorum quotannis obnoxia esset, Caleti Proprætor anno salutis 1428, de*

ratione litis judiciario ordine se instruens, id, sicut annotatum a me est, comperisse judicavit." —Robert Gaguin, lib. II. fol. 17.

[46] *Mémoires de l'Académie des Inscriptions, IV. p. 728.— The question is also discussed in the Traité de la Noblesse, by M. de la Roque; in the Mercure de France, for January, 1726; and in a Latin treatise by Charles Malingre, entitled "De falsâ regni Yvetoti narratione, ex majoribus commentariis fragmentum."*

[47] *Précis Analytique des Travaux de l'Académie de Rouen, 1811, p. 181.*

LETTER VII
ON THE STATE OF AFFAIRS IN FRANCE

(*Rouen, June,* 1818.)

Abandoning, for the present, all discussion of the themes of the elder day, I shall occupy myself with matters relating to the living world. The fatigued and hungry traveller, whose flesh is weaker than his spirit, is often too apt to think that his bed and his supper are of more immediate consequence than churches or castles. And to those who are in this predicament, there is a material improvement at Rouen, since I was last here: nothing could be worse than the inns of the year 1815; but four years of peace have effected a wonderful alteration, and nothing can now be better than the Hôtel de Normandie, where we have fixed our quarters. Objection may, indeed, be made to its situation, as to that of every other hôtel in the city; but this is of little moment in a town, where every house, whatever street or place it may front, opens into a court-yard, so that its views are confined to what passes within its own quadrangle; and, for excellence of accommodations, elegance of furniture, skill in cookery, civility of attendance, nay, even for what is more rare, neatness, our host, M. Trimolet, may challenge competition with almost any establishment in Europe. For the rent of the house, which is one of the most spacious in Rouen, he pays three thousand francs a year; and, as house-rent is one of the main standards of the value of the circulating medium, I will add, that our friend, M. Rondeau, for his, which is not only among the largest but among the most elegant and the best placed for business, pays but five hundred francs more. This, then, may be considered as the maximum at Rouen. Yet Rouen is far from being the place which should be selected by an Englishman, who retires to France for the purpose of economizing: living in general is scarcely one-fourth cheaper than in our own country. At Caen it is considerably more reasonable; on the banks of the Loire the expences of a family do not amount to one-half of the English cost; and still farther south a yet more sensible reduction takes place, the necessaries of life being cheaper by half than they are in Normandy, and house-rent by full four-fifths.

A foreigner can glean but little useful information respecting the actual state of a country through which he journeys with as much rapidity as I have done. And still less is he able to secern the truth from the falsehood, or to weigh the probabilities of conflicting testimony. I therefore originally intended to be silent on this subject. There is a story told, I believe, of Voltaire, at least it may be as well told of Voltaire as of any other wit, that, being once in company with a very talkative empty Frenchman, and a very glum and silent Englishman, he afterwards characterized them by saying, "l'un ne dit que des riens, et l'autre ne dit rien." Fearing that my political and statistical observations, which in good truth are very slender, might be ranked but too truly in the former category, I had resolved to confine them to my own notebook. Yet we all take so much interest in the destinies of our ancient rival and enemy, (I wish I could add, our modern friend,) that, according to my usual habit, I changed my determination within a minute after I had formed it; for I yielded to the impression, that even my scanty contribution would not be wholly unacceptable to you.

France, I am assured on all sides, is rapidly improving, and the government is satisfactory to all liberal men, in which number I include persons of every opinion, except the emigrants and those attached exclusively to the ancien régime. Men of the latter description are commonly known by the name of Ultras; and, speaking with a degree of freedom, which is practised here, to at least as great an extent as in England, they do not hesitate to express their decided disapprobation of the present system of government, and to declare, not only that Napoléon was more of a royalist than Louis, but that the King is a jacobin. They persuade themselves also, and would fain persuade others, that he is generally hated; and their doctrine is, that the nation is divided into three parties, ready to tear each other in pieces: the Ministerialists, who are few, and in every respect contemptible; the Ultras, not numerous, but headed by the Princes, and thus far of weight; and the Revolutionists, who, in point of numbers, as well as of talents and of opulence, considerably exceed the other two, and will, probably, ultimately prevail; so that these conflicts of opinion will terminate by decomposing the constitutional monarchy into a republic. To listen to these men, you might almost fancy they were quoting from Clarendon's History of the Rebellion in our own country; so entirely do their feelings coincide with those of the courtiers who attended Charles in his exile. Similar too is the reward they receive; for it is difficult for a monarch to be just, however he may in some cases he generous.

Yet even the Ultras admit that the revolution has been beneficial to France, though they are willing to confine its benefits to the establishment of the trial by jury, and the correction of certain abuses connected with the old system of nobility. Among the advantages obtained, they include the abolition of the game laws; and, indeed, I am persuaded, from all I hear, that this much-contested question could not receive a better solution than by appealing to the present laws in France. Game is here altogether the property of the land-owner; it is freely exposed for sale, like other articles of food; and every one is himself at liberty to sport, or to authorize his friend to do so over his property, with no other restriction than that of taking out a licence, or port d'armes, which, for fifteen francs, is granted without difficulty to any man of respectability, whatever may be his condition in life. In this particular, I cannot but think that France has set us an example well worthy of our imitation; and she also shews that it may be followed without danger; for neither do the pleasures of the field lose their relish, nor is the game extirpated. The former are a subject of conversation in almost every company; and, as to the latter, whatever slaughter may have taken place in the woods and preserves, at the first burst of the revolution, I am assured that a good sportsman may, at the present time, between Dieppe and Rouen kill with ease, in a day, fifty head of game, consisting principally of hares, quails, and partridges.

But, while these men thus restrict the benefits derived from the revolution, the case is far different with individuals of the other parties, all of whom are loud and unanimous in its praises. The good resulting from the republic has been purchased at a dreadful price, but the good remains; and those, who now enjoy the boon, are not inclined to remember the blood which drenched the three-colored banner. Thirty years have elapsed, and a new generation has arisen, to whom the horrors of the revolution live only in the page of history. But its advantages are daily felt in the equal nature and equal administration of the laws; in the suppression of the monasteries with their concomitant evils; in the restriction of the powers of the clergy; in the liberty afforded to all modes of religious worship; and in the abolition of all the edicts and mandates and prejudices, which secured to a peculiar sect and caste a monopoly of all the honors and distinctions of the common-wealth; for now, every individual of talent and character feels that the path to preferment and power is not obstructed by his birth or his opinions.

The constitutional charter, in its present state, is a subject of pride to the French, and a sure bulwark to the throne. The representative system is beginning to be generally appreciated, and particularly in commercial towns. The deputies of this department are to be changed the approaching autumn, and the minds of men are already anxiously bent upon selecting such representatives as may best understand and promote their local interests. Few acts of the Bourbon government have contributed more powerfully to promote the popularity of the King, than the law enacted in the course of last year, which abolished the double election, and enabled the voters to give their suffrages directly for their favorite candidate, thus putting a stop at once to a variety of unfair influence, previously exerted upon such occasions. The same law has also created a general interest upon the subject, never before known; the strongest proof of which is, that, of the six or eight thousand electors contained in this department, nearly the whole are expected now to vote, whereas not a third ever did so before. The qualifications for an elector and a deputy are uniform throughout the kingdom, and depending upon few requisites; nothing more being required in the former case, than the payment of three hundred francs per annum, in direct taxes, and the having attained the age of thirty; while an addition of ten years to the age, and the payment of one thousand francs, instead of three hundred, renders every individual qualified to be of the number of the elected. The system, however, is subject to a restriction, which provides, that at least one half of the representatives of each department shall be chosen from among those who reside in it.

In the beginning of the revolution, a much wider door was open: all that was then necessary to entitle a man to vote, was, that he should be twenty-one years of age, a Frenchman, and one who had lived for a year in the country on his own revenue, or on the produce of his labor, and was not in a state of servitude. It was then also decreed, that the electors should have each three livres a day during their mission, and should be allowed at the rate of one livre a league, for the distance from their usual place of residence, to that in which the election of members for their department is held. Such were the only conditions requisite for eligibility, either as elector or deputy; except, indeed, that the citizens in the primary assemblies, and the electors in the electoral assembly, swore that they would maintain liberty and equality, or die rather than violate their oath[48].

The wisdom and prudence of the subsequent alterations, few will be disposed to question: the system, in its present state, appears to me admirably qualified to attain the object in view; and such seems the general character of the French Constitutional Charter, which unites two excellent qualities, great clearness and great brevity. The whole is comprised in seventy-four short articles; and, that no Frenchman may plead ignorance of his rights or his duties, it is usually found prefixed to the almanacks. Some persons might, indeed, be inclined to deem this station as ominous; for, since the revolution began, the frame of the French government has sustained so many alterations, that, considering that several of their constitutions never outlived the current quarter, they may be fairly said to have had a new constitution in each year. How far the Bourbon charter will answer the purpose of serving as the basis of a code of laws for the government of an extensive kingdom, time only can determine. At present, it has the charm of novelty to recommend it; and there are few among us with whom novelty is not a strong attraction. Our friends on this side of the water are greatly belied, if it be not so with them.

The finances of the French municipalities are administered with a degree of fairness and attention, which might put many a body corporate, in a certain island, to the blush. Little is known in England respecting the administration of the French towns: the following particulars relating to the revenue and expences of Rouen, may, therefore, in some measure, serve as a scale, by which you may give a guess at the balance-sheet of cities of greater or lesser magnitude.—The budget amounted for the last year to one million two hundred thousand francs. The proposed items of expenditure must be particularized, and submitted to the Prefect and the Minister of the Interior, before they can be paid. In this sum is comprised the charge for the hospitals, which contain above three thousand persons, including foundlings, and for all the other public institutions, the number and excellence of which has long been the pride of Rouen. You must consider too, that every thing of this kind is, in France, national: individuals do nothing, neither is it expected of them; and herein consists one of the most essential differences between France and England. To meet this great expenditure, the city is provided with the rents of public lands, with wharfage, with tolls from the markets and the halles; and, above all, with the octroi, a tax that prevails through France, upon every article of consumption brought into the towns, and is collected at the barriers. The octroi, like turnpike-tolls or the post-horse duty with us, is farmed; two-thirds are received by the government, and the remaining one-third by the town. In Rouen it produced the last year one million four hundred and fifty thousand francs.—If, now, this sum appears to you comparatively greater than that of our large cities in England, you must recollect that, with us, towns are not liable to similar charges: our corporations support no museums, no academies, no learned bodies; and our infirmaries, and dispensaries, and hospitals, are indebted, as well for their existence as their future maintenance, to the piety of the dead, or the liberality of the living. Nor must we forget that, even in this great kingdom, Rouen, at present, holds the fifth place among the towns; though it was far from being thus, when Buonaparté, uniting the imperial to the iron crown, overshadowed with his eagle-wings the continent from the Baltic to Apulia; and when the mural crowns of Rome and Amsterdam stood beneath the shield of the "good city" of Paris.

The population of Rouen is estimated at eighty-seven thousand persons, of whom the greater number are engaged in the manufactories, which consist principally of cotton, linen, and woollen cloths, and are among the largest in France. At present, however, "trade is dull;" and hence, and as the politics of a trader invariably sympathize with his cash account, neither the peace, nor the English, nor the princes of the Bourbon dynasty, are popular here; for the articles manufactured at Rouen, being designed generally for exportation, ranged almost unrivalled over the continent, during the war, but now in every town they meet with competitors in the goods from England, which are at once of superior workmanship and cheaper. The latter advantage is owing very much to the greater perfection of our machinery, and, perhaps, still more to the abundance of coals, which enables us, at so small an expence, to keep our steam-engines in action, and thus to counterbalance the disproportion in the charge of manual labor, as well as the many disadvantages arising from the pressure of our heavy taxation.— But I must cease. An English fit of growling is coming upon me; and I find that the Blue Devils, which haunt St. Stephen's chapel, are pursuing me over the channel.

Footnotes:

[48] *Moore's Journal of a Residence in France, I. p. 82.*

LETTER VIII
MILITARY ANTIQUITIES—LE VIEUX CHÂTEAU—ORIGINAL PALACE OF THE NORMAN DUKES—HALLES OF ROUEN—MIRACLE AND PRIVILEGE OF ST. ROMAIN—CHÂTEAU DU VIEUX PALAIS—PETIT CHÂTEAU—FORT ON MONT STE. CATHERINE—PRIORY THERE—CHAPEL OF ST. MICHAEL—DEVOTEE

(Rouen, June, 1818)

My researches in this city after the remains of architectural antiquity of the earlier Norman æra, have hitherto, I own, been attended with little success. I may even go so far as to say, that I have seen nothing in the circular style, for which it would not be easy to find a parallel in most of the large towns in England. On the other hand, the perfection and beauty of the specimens of the pointed style, have equally surprised and delighted me. I will endeavor, however, to take each object in its order, premising that I have been materially assisted in my investigations by M. Le Prevost and M. Rondeau, but especially by the former, one of the most learned antiquaries of Normandy.

Of the fortifications and castellated buildings in Rouen very little indeed is left[49], and that little is altogether insignificant; being confined to some fragments of the walls scattered here and there[50], and to three circular towers of the plainest construction, the remains of the old castle, built by Philip Augustus in 1204, near to the Porte Bouvreuil, and hence commonly known by the name of the Château de Bouvreuil or le Vieux Château.—It is to the leading part which this city has acted in the history of France, that we must attribute the repeated erection and demolition of its fortifications.

An important event was commemorated by the erection of the *old castle,* it having been built upon the final annexation of Normandy to the crown of France, in consequence of the weakness of our ill-starred monarch,—John Lackland. The French King seems to have suspected that the citizens retained their fealty to their former sovereign. He intended that his fortress should command and bridle the city, instead of defending it. The town-walls were razed, and the *Vieille Tour*, the ancient palace of the Norman Dukes, levelled with the ground.—But, as the poet says of language, so it is with castles,—

... "mortalia facta peribunt,

Nec castellorum stet honos et gratia vivax;"

and, in 1590, the fortress raised by Philip Augustus experienced the fate of its predecessors; it was then ruined and dismantled, and the portion which was allowed to stand, was degraded into a jail. Now the three[51] towers just mentioned are alone remaining, and these would attract little notice, were it not that one of them bears the name of the Tour de la Pucelle, as having been, in 1430, the place of confinement of the unfortunate Joan of Arc, when she was captured before Compiégne and brought prisoner to Rouen.

It must be stated, however, that the first castle recorded to have existed at Rouen, was built by Rollo, shortly after he had made himself master of Neustria. Its very name is now lost; and all we know concerning it is, that it stood near the quay, at the northern extremity of the town, in the situation subsequently occupied by the Church of St. Pierre du Châtel, and the adjoining monastery of the Cordeliers.

After a lapse of less than fifty years, Rouen saw rising within her walls a second castle, the work of Duke Richard Ist, and long the residence of the Norman sovereigns. This, from a tower of great strength which formed a part of it, and which was not demolished till the year 1204, acquired the appellation of *la Vieille Tour*; and the name remains to this day, though the building has disappeared.

The space formerly occupied by the scite of it is now covered by the halles, considered the finest in France. The historians of Rouen, in the usual strain of hyperbole, hint that their halles are even the finest in the world[52], though they are very inferior to their prototypes at Bruges and Ypres. The hall, or exchange, allotted to the mercers, is two hundred and seventy-two feet in length, by fifty feet wide: those for the drapers and for wool are, each of them, two hundred feet long; and all these are surpassed in size by the corn-hall, whose length extends to three hundred feet. They are built round a large square, the centre of which is occupied by numberless dealers in pottery, old clothes, &c.; and, as the day on which we chanced to visit them was a Friday, when alone they are opened for public business, we found a most lively, curious, and interesting scene.

It was on the top of a stone staircase, the present entry to the halles, that the annual ceremony[53] of delivering and pardoning a criminal for the sake of St. Romain, the tutelary protector of Rouen, was performed on Ascension-day, according to a privilege exercised, from time immemorial, by the Chapter of the Cathedral.

The legend is romantic; and it acquires a species of historical importance, as it became the foundation of a right, asserted even in our own days. My account of it is taken from Dom Pommeraye's History of the Life of the Prelate[54].—He has been relating many miracles performed by him, and, among others, that of causing the Seine, at the time of a great inundation, to retire to its channel by his command, agreeably to the following beautiful stanza of Santeuil:—

> "Tangit exundans aqua civitatem;
>
> Voce Romanus jubet efficaci;
>
> Audiunt fluctus, docilisque cedit
>
> Unda jubenti."

Our learned Benedictine thus proceeds:—"But the following miracle was deemed a far greater marvel, and it increased the veneration of the people towards St. Romain to such a degree, that they henceforth regarded him as an actual apostle, who, from the authority of his office, the excellence of his doctrine, his extreme sanctity, and the gift of miracles, deserved to be classed with the earliest preachers of our holy faith. In a marshy spot, near Rouen, was bred a dragon, the very counterpart of that destroyed by St. Nicaise. It committed frightful ravages; lay in wait for man and beast, whom it devoured without mercy; the air was poisoned by its pestilential breath, and it was alone the cause of greater mischief and alarm, than could have been occasioned by a whole army of enemies. The inhabitants, wearied out by many years of suffering, implored the aid of St. Romain; and the charitable and generous pastor, who dreaded nothing in behalf of his flock, comforted them with the assurance of a speedy deliverance. The design itself was noble; still more so was the manner by which he put it in force; for he would not be satisfied with merely killing the monster, but undertook also to bring it to public execution, by way of atonement for its cruelties. For this purpose, it was necessary that the dragon should be caught; but when the prelate required a companion in the attempt, the hearts of all men failed them. He applied, therefore, to a criminal condemned to death for murder; and, by the promise of a pardon, bought his assistance, which the certain prospect of a scaffold, had he refused to accompany the saint, caused him the more willingly to lend. Together they went, and had no sooner reached the marsh, the monster's haunt, than St. Romain, approaching courageously, made the sign of the cross, and at once put it out of the power of the dragon to attempt to do him injury. He then tied his stole around his neck, and, in that state, delivered him to the prisoner, who dragged him to the city, where he was burned in the presence of all the people, and his ashes thrown into the river.—The manuscript of the Abbey of Hautmont, from which this legend is extracted, adds, that such was the fame of this miracle throughout France, that Dagobert, the reigning sovereign, sent for St. Romain to court, to hear a true narrative of the fact from his own lips; and, impressed with reverent awe, bestowed the celebrated privilege upon him and his successors for ever."

The right has, in comparatively modern times, been more than once contested, but always maintained; and so great was the celebrity of the ceremony, that princes and potentates have repeatedly travelled to Rouen, for the purpose of witnessing it. There are not wanting, however, those[55] who treat the whole story as allegorical, and believe it to be nothing more than a symbolical representation of the subversion of idolatry, or of the confining of the Seine to its channel; the winding course of the river being typified by a serpent, and the word Gargouille corrupted from gurges. Other writers differ in minor points of the story, and alledge that the saint had two fellow adventurers, a thief as well as a murderer, and that the former ran away, while the latter stood firm. You will see it thus figured in

a modern painting on St. Romain's altar, in the cathedral; and there are two persons also with him, in the only ancient representation of the subject I am acquainted with, a bas-relief which till lately existed at the Porte Bouvreuil, and of which, by the kindness of M. Riaux, I am enabled to send you a drawing.

To keep alive the tradition, in which Popish superstition has contrived to blend Judaic customs with heathen mythology, the practice was, that the prisoner selected for pardon should be brought to this place, called the chapel of St. Romain, and should here be received by the clergy in full robes, headed by the archbishop, and bearing all the relics of the church; among others, the shrine of St. Romain, which the criminal, after having been reprimanded and absolved, but still kneeling, thrice lifted, among the shouts of the populace, and then, with a garland upon his head and the shrine in his hands, accompanied the clergy in procession to the cathedral[56].—But the revolution happily consigned the relics to their kindred dust, and put an end to a privilege eminently liable to abuse, from the circumstance of the pardon being extended, not only to the criminal himself, but to all his accomplices; so that, an inferior culprit sometimes surrendered himself to justice, in confidence of interest being made to obtain him the shrine, and thus to shield under his protection more powerful and more guilty delinquents. The various modifications, however, of latter times, had so abridged its power, that it was at last only able to rescue a man guilty of involuntary homicide[57]. We may hope, therefore, it was not altogether deserving the hard terms bestowed upon it by Millin[58] who calls it the most absurd, most infamous, and most detestable of all privileges, and adduces a very flagrant instance of injustice committed under its plea.—D'Alégre, governor of Gisors, in consequence of a private pique against the Baron du Hallot, lord of the neighboring town of Vernon, treacherously assassinated him at his own house, while he was yet upon crutches, in consequence of the wounds received at the siege of Rouen. This happened during the civil wars; in the course of which, Hallot had signalized himself as a faithful servant, and useful assistant to the monarch. The murderer knew that there were no hopes for him of royal mercy; and, after having passed some time in concealment and as a soldier in the army of the league, he had recourse to the Chapter of the Cathedral of Rouen, from whom he obtained the promise of the shrine of St. Romain. To put full confidence, however, even in this, would, under such circumstances, have been imprudent. The clergy might break their word, or a mightier power might interpose. D'Alégre, therefore, persuaded a young mam, formerly a page of his, of the name of Pehu, to surrender himself as guilty of the crime; and to him the privilege was granted; under the sanction of which, the real culprit, and several of his accomplices in the assassination, obtained a free pardon. The widow and daughter of Hallot, in vain remonstrated: the utmost that could be done, after a tedious law-suit, was to procure a small fine to be imposed upon Pehu, and to cause him to be banished from Normandy and Picardy and the vicinity of Paris. But regulations were in consequence adopted with respect to the exercise of the privilege; and the pardons granted under favor of it were ever afterwards obliged to be ratified under the high seal of the kingdom.

The Château du Vieux Palais and le petit Château like the edifices which I have already noticed, have equally yielded to time and violence. M. Carpentier has furnished us with representations of both these castles, drawn and etched by himself, in the Itinerary of Rouen. The first of them has also been inaccurately figured by Ducarel, and satisfactorily by Millin, in the second volume of his Antiquités Nationales; where, to the pen of this most meritorious and indefatigable writer, of whom, as of our Goldsmith, it may be justly said, that "nullum ferè scribendi genus non tetigit, nullum quod tetigit non ornavit," it affords materials for a curious memoir, blended with the history of our own Henry Vth, and of Henry IVth, of France. The castle was the work of the first of these sovereigns, and was begun by him in 1420, two years after a seven months' siege had put him in possession of the city, long the capital of his ancestors, and had thus rendered him undisputed master of Normandy. This was an event worthy of being immortalised; and it may easily be imagined that private feelings had no little share in urging him to erect a magnificent palace, intended at once as a safeguard for the town, and a residence for himself and his posterity. The right to build it was an express article in the capitulation he granted to Rouen, a capitulation of extreme severity[59], and purchased at the price of three hundred thousand golden crowns, as well as of the lives of three of the most distinguished citizens; Robert Livret, grand-vicar of the archbishop, John Jourdain, commander of the artillery, and Louis Blanchard, captain of the train-bands. The two first of these were, however, suffered to ransome themselves; the last, a man of distinguished honor and courage, was beheaded; but Henry, much to his credit, made no farther use of his victory, and even consented to pay for the ground required for his castle. He selected for the purpose, the situation where, defence was most needed, upon the extremity of the quay, by the side of the river, near the entrance from Dieppe and Havre. A row of handsome houses now fills the chief part of the space occupied by the building, which, at a subsequent period, was again connected with English history[60], as the residence of our James IInd, after the battle of La Hague; before his spirit was yet sufficiently broken to suffer him to give up all thoughts of the British crown, and to accept the asylum offered by Louis XIVth, in the obscure tranquillity of Saint Germain's. It continued perfect till the time of the revolution, and was of great extent and strength, defended by massy circular towers, surrounded by a moat, and approachable only by a draw-bridge.

The castle, which still remains to be described, and whose smaller size is sufficiently denoted by its name, was also built by the same monarch, but it was raised upon the ruins of a similar edifice that had existed since the days of King John. Being situated at the foot of the bridge, the older castle had been selected as the spot where it was stipulated that the soldiers, composing the Anglo-Norman garrison, should lay down their arms, when the town surrendered to Philip Augustus.—It was known from very early time by the appellation of the Barbican, a term of much disputed signification as well as origin: if we are to conclude, according to some authorities, that it denoted either a mere breast-work, or a watch-tower, or an appendage to a more important fortress, it would appear but ill applied to a building like the one in question. I should rather believe it designated an out-post of any kind; and I would support my conjecture by this very castle, which was neither upon elevated ground, nor dependent on any other. It consisted of two square edifices, similar to what are called the pavillions of the Thuilleries, flanked by small circular towers with conical roofs, and connected by an embattled wall. Not more than fifty years have passed since its demolition; yet no traces of it are to be found.

A few rocky fragments, appearing now to bid defiance to time, indicate the scite of the fortress, which once arose on the summit of Mont Ste. Catherine, and which, though dismantled by Henry IVth, and reduced to a state of dilapidation, was still suffered to maintain its ruined existence till a few years ago. Its commanding situation, upon an eminence three hundred and eighty feet high and immediately overhanging the city, could not but render it of great importance towards the defence of the place; and we accordingly find that Taillepied, who probably wrote before its demolition, gives it as his opinion, that whoever is in possession of Mont Ste. Catherine, is also master of the town, if he can but have abundant supplies of water and provisions;—no needless stipulation! At the same time, it must be admitted that the fort was equally liable to be converted into the means of annoyance. Such actually proved the case in 1562, at which time it was seized by the Huguenots; and considerations of this nature most probably prevailed with the citizens, when they declined the offer made by Francis Ist, who proposed at a public meeting to enlarge the tower into an impregnable citadel. In the hands of the Protestants, the fortress, such as it was, proved sufficient to resist the whole army of Charles IXth, during several days.— Rouen was stoutly defended by the reformed, well aware of the sanguinary dispositions of the bigotted monarch. They yielded, and he sullied his victory by giving the city up to plunder, during twenty-four hours; and we are told, that it was upon this occasion he first tasted heretical blood, with which, five years afterwards, he so cruelly gorged himself on the day of St. Bartholomew. Catherine of Medicis accompanied him to the siege; and it is related that she herself led him to the ditches of the ramparts, in which many of their adversaries had been buried, and caused the bodies to be dug up in his presence, that he might be accustomed to look without horror upon the corpse of a Protestant!

Near the fort stood a priory[61], whose foundation is dated as far back as the eleventh century, when Gosselin, Viscount of Rouen, Lord of Arques and Dieppe, having no son to inherit his wealth, was induced to dispose of it "to pious uses," by the persuasions of two monks, who had wandered in pilgrimage from the monastery of Saint Catherine, on Mount Sinai. These good men assured him, that, if he dedicated a church to the martyred daughter of the King of Alexandria, the stones employed in building it would one day serve him as so many stepping-stones to heaven. They confirmed him in his resolution, by presenting him with one of the fingers of Saint Catherine. To her, therefore, the edifice was made sacred, and hence it is believed that the hill also took its name. In the Golden Legend, we find an account of the translation of the finger to Rouen not wholly reconcileable with this history.—According to the veracious authority of James of Voragine, there were certain monks of Rouen, who journeyed even until the Arabian mountain. For seven long years did they pray before the shrine of the Queen Virgin and Martyr, and also did they implore her to vouchsafe to grant them some token of her favor; and, at length, one of her fingers suddenly disjointed itself from the dead hand of the corpse.—"This gift," as the legend tells, "they received devoutly, and with it they returned to their monastery at Rouen."—Never was a miracle less miraculous; and it is fortunately now of little consequence to inquire whether the mouldering relic enriched an older monastery, or assisted in bestowing sanctity on a rising community. According to the pseudo-hagiologists, the corpse of Saint Catherine was borne through the air by angels, and deposited on the summit of Mount Sinai, on the spot where her church is yet standing. Conforming, as it were, to the example of the angels, it was usual, in the middle ages, to erect her religious buildings on an eminence. Various instances may be given of this practice in England, as well as in France: such is the case near Winchester, near Christ-Church, in the Isle of Wight, and in many other places. St. Michael contested the honor with her; and he likewise has a chapel here, whose walls are yet standing. Its antiquity was still greater than that of the neighboring monastery; a charter from Duke Richard IInd, dated 996, speaking of it as having had existence before his time, and confirming the donation of it to the Abbey of St. Ouen. But St. Michael's never rivalled the opulence of Saint Catherine's priory.—Gosselin himself, and Emmeline his wife, lay buried in the church of the latter, which is said to have been large, and to have resembled in its structure that of St. Georges de Bocherville: it is also recorded, that it was ornamented with many beautiful paintings; and loud praises are bestowed upon its fine peal of bells. The epitaph of the founder speaks of him, as—

> "Premier Autheur des mesures et poids
>
> Selon raison en ce päis Normand."

It is somewhat remarkable, that there appear to have been only two other monumental inscriptions in the church, and both of them in memory of cooks of the convent; a presumptive proof that the holy fathers were not inattentive to the good things of this world, in the midst of their concern for those of the next.—The first of them was for Stephen de Saumere,—

> "Qui en son vivant cuisinier
>
> Fut de Révérend Pere en Dieu,
>
> De la Barre, Abbé de ce lieu."

The other was for—

> "Thierry Gueroult, en broche et en fossets
>
> Gueu très-expert pour les Religieux."

The fort and the religious buildings all perished nearly at the same time: the former was destroyed at the request of the inhabitants, to whom Henry IVth returned on that occasion his well-known answer, that he "wished for no other fortress than the hearts of his subjects;" the latter to gratify the avarice of individuals, who cloked their true designs under the plea that the buildings might serve as a harbor for the disaffected.

Of the origin of the fort I find no record in history, except what Noel says[62], that it appears to have been raised by the English while they were masters of Normandy; but what I observed of the structure of the walls, in 1815, would induce me to refer it without much hesitation to the time of the Romans. Its bricks are of the same form and texture as those used by them; and they were ranged in alternate courses with flints, as is the case at Burgh Castle, at Richborough, and other Roman edifices in England. That the fort was of great size and strength is sufficiently shewn by the depth, width, and extent of the entrenchments still left, which, particularly towards the plain, are immense; and, if credence may be given to common report, in such matters always apt to exaggerate, the subterraneous passages indicate a fortress of importance.

It chanced, that I visited the hill on Michaelmas-day, and a curious proof was afforded me, that, at however low an ebb religion may be in France, enthusiastic fanaticism is far from extinct. A man of the lower classes of society was praying before a broken cross, near St. Michael's Chapel, where, before the revolution, the monks of St. Ouen used annually on this day to perform mass, and many persons of extraordinary piety were wont to assemble the first Wednesday of every month to pray and to preach, in honor of the guardian angels. His manner was earnest in the extreme; his eyes wandered strangely; his gestures were extravagant, and tears rolled in profusion down a face, whose every feature bore the strongest marks of a decided devotee. A shower which came at the moment compelled us both to seek shelter within the walls of the chapel, and we soon became social and entered into conversation. The ruined state of the building was his first and favorite topic: he lamented its destruction; he mourned over the state of the times which could countenance such impiety; and gradually, while he turned over the leaves of the prayer-book in his hand, he was led to read aloud the hundred and thirty-sixth psalm, commenting upon every verse as he proceeded, and weeping more and more bitterly, when he came to the part commemorating the ruin of Jerusalem, which he applied, naturally enough, to the captive state of France, smarting as she then was under the iron rod of Prussia. Of the other allies, including even the Russians, he owned that there was no complaint to be made: "they conduct themselves," said he, "agreeably to the maxim of warfare, which says 'battez-vous contre ceux qui vous opposent; mais ayez pitié des vaincus.' Not so the Prussians: with them it is 'frappez-çà, frappez-là, et quand ils entrent dans quelque endroit, ils disent, il nous faut çà, il nous faut là, et ils le prennent d'autorité.' Cruel Babylon!"—"Yet, even admitting all this," we asked, "how can you reconcile with the spirit of christianity the permission given to the Jews by the psalmist, to 'take up her little ones and dash them against the stones.'"—"Ah! you misunderstand the sense, the psalm does not authorize cruelty;—mais, attendez! ce n'est pas ainsi: ces pierres là sont Saint Pierre; et heureux celui qui les attachera à Saint Pierre; qui montrera de l'attachement, de l'intrépidité pour sa religion."—Then again, looking at the chapel, with tears and sobs, "how can we expect to prosper, how to escape these miseries, after having committed such enormities?"—His name, he told us, was Jacquemet, and my companion kindly made a sketch of his face, while I noted down his words.

This specimen will give you some idea of the extraordinary influence of the Roman catholic faith over the mind, and of the curious perversions under which it does not scruple to take refuge.

Leaving for the present the dusty legends of superstition, I describe with pleasure my recollections of the glorious prospect over which the eye ranges from the hill of Saint Catherine.—The Seine, broad, winding, and full of islands, is the principal feature of the landscape. This river is distinguished by its sinuosity and the number of islets which it embraces, and it retains this character even to Paris. Its smooth tranquillity well contrasts with the life that is imparted to the scene, by the shipping and the bustle of the quays. The city itself, with its verdant walks, its spacious manufactories, its strange and picturesque buildings, and the numerous spires and towers of its churches, many of them in ruins, but not the less interesting on account of their decay, presents a foreground diversified with endless variety of form and color. The bridge of boats seems immediately at our feet; the middle distance is composed of a plain, chiefly consisting of the richest meadows, interspersed copiously with country seats and villages embosomed in wood; and the horizon melts into an undulating line of remote hills.

Footnotes:

[49] Farin, Histoire de Rouen, I. p. 97.

[50] In a paper printed in the Transactions of the Rouen Academy for 1818, p. 177, it appears that, so late as 1789, a considerable portion of very old walls was discovered under-ground; and that they consisted very much of Roman bricks. Among them was also found a Roman urn, and eighty or more medals of the same nation, but none of them older than Antoninus.—From this it appears certain that Rouen was a Roman station, though of its early history we have no distinct knowledge.

[51] These are the Tour du Gascon, Tour du Donjon, and Tour de la Pucelle.

[52] Histoire de Rouen, I. p. 32.

[53] Histoire de Rouen, III. p. 34.

[54] It is also worth while to read the following details from Bourgueville, (Antiquités de Caen, p. 33) whose testimony, as that of an eye-witness to much of what he relates, is valuable:—"Ils ont le Privilege Saint Romain en la ville de Rouen et Eglise Cathédrale

du lieu, au iour de l'Ascension nostre Seigneur de deliurer un prisonnier, qui leur fut concedé par le Roy d'Agobert en memoire d'un miracle que Dieu fist par saint Romain Archeuesque du lieu, d'auoir deliuré les habitans d'un Dragon qui leur nuisoit en la forest de Rouuray pres ladite ville: pour lequel vaincre il demanda à la justice deux prisonniers dignes de mort, l'un meurtrier et l'autre larron: le larron eut si grand frayeur qu'il s'enfuit, et le meurtrier demeura auecque ce saint homme qui vainquit ce Serpent. C'est pourquoy l'on dit encore en commun prouerbe, il est asseuré comme vn meurtrier. Ce privilege de deliurance ne doit estre accordé aux larrons.—Saint Ouen successeur de S. Romain, Chancelier dudit Roy d'Agobert viron l'an 655, impetra ce priuilege: dont ie n'en deduiray en plus oultre les causes, pour ce qu'elles sont assez communes et notoires, et feray seulement cest aduertissement, qu'il y a danger que messieurs les Ecclesiastiques le perdent, acause qu il s'y commet le plus souuent des abus, par ce qu'il se doit donner en cas pitoyable et non par authorité ou faueurs de seigneurs, comme aussi ne se doit estendre, sinon à ceux qui sont trouuez actuellement prisonniers sans fraude, et non à ceux qui s'y rendent le soir precedent comme estans asseurez d'obtenir ce priuilege, combien qu'ils ayent commis tous crimes execrables et indignes d'un tel pardon, voire et que les Ecclesiastiques n'ayent eu loisir d'avoir veu et bien examinez leur procez. Aussi ce beau priuilege est enfraint en ce que ceux qui l'obtiennent doiuent assister par sept annees suiuantes aux processions au tour de la Fierte S. Romain, portant vne torche ardante selon qu'il leur est chargé faire. Ce qui est de ceste heure trop contemné: et tel mespris leur pourroit estre reproché comme indignes et contempteurs d'vn tel pardon. Vn surnommé Saugrence pour auoir abusé d'un tel priuilege fut quelque temps apres retrudé et puni de la peine de la roüe pour auoir confesse des meurtres en agression pour sauuer aucuns nobles ou nocibles qui les auoient commis.—Il s'est faict autres fois et encore du temps de ma ieunesse de grands festins, danses, mommeries ou mascarades audit iour de l'Ascension, tant par les feturiers de ceste confrairie saint Romain que autres ieunes hommes auec excessiues despences: et s'appelloit lors tel iour Rouuoysons, à cause que les processions rouent de lieu en autre, et disoit l'on comme en prouerbe, quand aucuns desbauchez declinoient de biens qu'ils auoient fait Rouuoysons, à sçauoir perdu leurs biens en trop uoluptueuses despenses et mommeries sur chariots, qui se faisoient de nuict par les ruës quelque saison d'Esté qu'il fust, pour plus grandes magnificences."

[55] *See Gallia Christiana, XI. p. 12.*

[56] *A minute and very curious account of the whole of this ceremony, from the first claiming of the prisoner to his final deliverance, is given in Tuillepied's Antiquités de Rouen, p. 79.*

[57] *Noel, Essais sur le Département de la Seine Inférieure, II. p. 228.*

[58] *Antiquités Nationales, II. No. 21 p. 3*

[59] *Millin, Antiquités Nationales, II. No. 20. p. 3.*

[60] *Noel, Essais sur le Département de la Seine Inférieure, II. p. 209*

[61] *Farin, Histoire de Rouen, V. p. 113.*

[62] *Essais sur le Département de la Seine Inférieure, II. p. 210.*

LETTER IX
ANCIENT ECCLESIASTICAL ARCHITECTURE— CHURCHES OF ST. PAUL AND ST. GERVAIS— HOSPITAL OF ST. JULIEN—CHURCHES OF LERY, PAVILLY, AND YAINVILLE

(Rouen, June, 1818.)

We, *East Angles*, are accustomed to admire the remains of Norman architecture, which, in our counties, are perhaps more numerous and singular than in any other tract in England. The noble castle of Blanchefleur still honors our provincial metropolis, and although devouring eld hath impaired her charms and converted her into a very dusky beauty, the fretted walls still possess an air of antique magnificence which we seek in vain when we contemplate the towers of Julius or the frowning dungeons of Gundulph. Our cathedral retains the pristine character which was given to the edifice, when the Norman prelate abandoned the seat of the Saxon bishop, and commanded the Saxon clerks to migrate into the city protected or inclosed by the garrison of his cognate conquerors. Even our villages abound with these monuments. The humbler, though not less sacred structures in which the voice of prayer and praise has been heard during so many generations, equally bear witness to Norman art, and, I may say, to Norman piety; and when we enter the sheltered porch, we behold the fantastic sculpture and varied foliage, encircling the arch which arose when our land was ruled by the Norman dynasty.

Comparatively speaking, Rouen is barren indeed of such relics. Its military antiquities are swept away; and the only specimens of early ecclesiastical architecture are found in the churches of St. Paul and St. Gervais, both of them, in themselves, unimportant buildings, and both so disfigured by subsequent alterations, that they might easily escape the notice of any but an experienced eye. Of these, the first is situated by the side of the road to Paris, under Mont Ste. Catherine, yet, still upon an eminence, beneath which are some mineral springs, that were long famous for their medicinal qualities, but have of late years been abandoned, and the spa-drinkers now resort to others in the quarter of the town called de la

Maréquerie. Both the one and the other are highly ferruginous, but the latter most strongly impregnated with iron.

The chancel is the only ancient part of the present church of St. Paul's, and even this must be comparatively modern, if any confidence may be placed in the current tradition, that the building, in its original state, was a temple of Adonis or of Venus, to both which divinities the early inhabitants of Rouen are reported to have paid peculiar homage. They were worshipped in vice and impurity[63]; nor were the votaries deterred by the evil spirits who haunted the immediate vicinity of the temple, and who gave rise to so fetid and infectious a vapor, that it often proved fatal! This very remark seems to indicate the scite of the church of St. Paul, with its neighboring sulphureous waters. St. Romain demolished the temple, and dispersed the sinners. Farin, in his History of Rouen[64], says, that the church was repeatedly destroyed and rebuilt by the Norman Dukes, to some of whom, the chancel, which is now standing, probably owes its existence. The nave is evidently of much more modern construction: it is thrice the width of the other part, from which it is separated by a circular arch. The eastern extremity differs from that of any other church I ever saw in Normandy or in England: it ends in three circular compartments, the central considerably the largest and most prominent, and divided from the others, which serve as aisles, by double arches, a larger and smaller being united together. This triple circular ending is, however, only observable without; for, in the interior, the southern part has been separated and used as a sacristy; the northern is a lumber-room. In the latter division, M. le Prevost desired us to notice a piece of sculpture, so covered with dirt and dust that it could scarcely be seen, but evidently of Roman workmanship, and, probably, of the fourth century, if we may judge from its resemblance to some ornaments[65] upon the pedestal of the obelisk raised by Theodosius, in the Hippodrome of Constantinople. Our friend's conjecture is, that it had originally served for an altar: perhaps

Sculpture in the Church of St. Paul at Rouen.

it might, with equal probability, be supposed to have been a tomb.—The corbels on the exterior of this building are strange and fanciful.

St. Gervais also stands without the walls of Rouen; but at the opposite end of the town, upon a hill adjoining the Roman road to Lillebonne, and near the Mont aux Malades, a place so called, as having been selected in the eleventh century, on account of the salubrity of its air, for the situation of a monastery, destined for the reception of lepers. Upon this eminence, the Norman Dukes had likewise originally a palace; and, it was to this, that William the Conqueror caused himself to be conveyed, when attacked with his mortal illness, after having wantonly reduced the town of Mantes to ashes. Here, too, this mighty monarch breathed his last, and left a sad warning to future conquerors, deserted by his friends and physicians the moment he was no more; while his menials plundered his property, and his body lay naked and neglected in the hall[66].

The ducal palace, and the monastic buildings of the priory, once connected with it, are now completely destroyed. Fortunately, however, the church still remains, though parochial and in poverty. It preserves some portions of the original structure, more interesting from their features than their extent. The exterior of the apsis is very curious: it is obtusely angular, and faced at the corners with large rude columns, of whose capitals some are Doric or Corinthian, others as wild as the fancies of the Norman lords of the country. None reach so high as the cornice of the roof, it having been the intention of the original architect, that a portion of work should intervene between the summit of the capitals and this member. A capital to the north is remarkable for the eagles carved upon it, as if with some allusion to Roman power. But the most singular part of this church is the crypt under the apsis, a room about thirty feet long by fourteen wide, and sixteen high, of extreme simplicity, and remote antiquity. Round it runs a plain stone bench; and it is divided into two unequal parts by a circular arch, devoid of columns or of any ornament whatever, but disclosing, in the composition of its piers, Roman bricks and other débris, some of them rudely sculptured. Here, according to Ordericus Vitalis[67], was interred the body of St. Mellonus, the first Archbishop of Rouen, and one of the apostles of Neustria; and here, his tomb, and that of his successor, Avitien, are shewn to this day, in plain niches, on opposite sides of the wall. St. Mello's remains however, were not suffered to rest in peace; for, about five hundred and seventy years after his death, which happened in the year 314, they were removed to the castle of Pontoise, lest the canonized corpse should be violated by the heathen Normans. In the diocese of Rouen St. Mello is honored with particular veneration; and the history of the prelates of the see contains many curious, and not unedifying stories of the miracles he performed. His feast, together with that of St. Nicasius, his companion, is celebrated on the second of October; and their labors are commemorated with a hymn appointed for their festival: —

"Primæ vos canimus gentis apostolos,
 Per quos relligio tradita patribus;
 Errorisque jugo libera Neustria
 CHRISTO sub duce militat.

"Facti sponte suis finibus exules
 Hùc de Romuleis sedibus advolant;
 Merces est operis, si nova consecrent
 Vero pectora Numini.

"Qui se pro populis devovet hostiam
 Mellonus tacitâ se nece conficit;
 Mactatus celeri morte Nicasius
 Christum sanguine prædicat."

Heretics as we are, we ought not to refrain from respecting the zeal even of a saint of the Catholic calendar, when thus exerted. Besides which, he has another claim upon our attention: our own island gave him birth, and he appeared at Rome as the bearer of the annual tribute of the Britons, at the very time when he was converted to Christianity, whose light he had afterwards the glory of diffusing over Neustria. The existence of these tombs and the antiquity of the crypt, recorded as it is by history and confirmed by the style of its architecture, have given currency to the tradition, which points it out as the only temple where the primitive Christians of Neustria dared to assemble for the performance of divine service. Many stone coffins have also been discovered in the vicinity of the church. These sarcophagi seem to confirm the general tradition: they are of the simplest form, and apparently as ancient as the crypt, and they were so placed in the ground that the heads of the corpses were turned to the east, a position denoting that the dead received Christian burial.

Another opportunity will be afforded me of speaking of the church of St. Ouen; but, as a singular relic of Norman architecture, I must here notice the round tower on the south side of the choir, probably part of the original edifice, finished by the Abbot, William Balot, and dedicated by the Archbishop Géoffroi, in 1126. It consists of two stories, divided by a billetted moulding. Respecting its use it would not now be easy to offer a probable conjecture: the history of the abbey, indeed, mentions it under the title of la Chambre des Clercs, and supposes that it was formerly a chapel[68]; but its shape and size do not seem to confirm that opinion.

The chapel of the suppressed lazar-house of St. Julien, situated about three miles from Rouen, on the opposite side of the Seine, is more perfect than either St. Paul or St. Gervais, and, consequently, more valuable to the architect. This building, without spire or tower, and divided into three parts of unequal length and height, the nave, the choir, and the circular apsis, externally resembles one of the meanest of our parish-churches, such as a stranger, judging only from the exterior, would be almost equally likely to consider as a place of worship, or as a barn. It is, however, if I am not mistaken, one of the purest and most perfect specimens of the Norman æra. I know of no building in England, which resembles it so nearly as the chancel of Hales Church, in Norfolk; but the latter has been exposed to material alterations, while the chapel of which I am speaking is externally quite regular in its design, being divided throughout its whole length into small compartments, by a row of shallow buttresses rising from the ground to the eaves of the roof, without any partition into splays. Those on the south side are still in their primæval state; but a buttress of a subsequent, though not recent, date, has been built up against almost every one of the original buttresses on the north side, by way of support to the edifice. Each division contains a single narrow circular-headed window: beneath these is a plain moulding, continued uninterruptedly over the buttresses as well as the wall, thus proving both to be coeval; another plain moulding runs nearly on a level with the tops of the windows, and takes the same circular form; but it is confined to the spaces between the buttresses. There are no others. The entrance was by circular-headed doors at the west end and south side, both of them very plain; but particularly the latter. The few ornaments of the western are as perfect and as sharp as if the whole were the work of yesterday. This part of the church has, however, been exposed to considerable injury, owing to its having joined the conventual buildings, which were destroyed at the revolution. The inside is, like the exterior, almost perfect, but it is very much more rich, uniting to the common ornaments of Norman architecture, capitals, in some instances, of classical beauty. The ceiling is covered with paintings of scriptural subjects, which still remain, notwithstanding that the building is now desecrated, and used as a woodhouse by the neighboring farmer.

The date of the erection of the chapel is well ascertained[69]. The hospital was founded in 1183, by Henry Plantagenet, as a priory for the reception of unmarried ladies of noble blood, who were destined for a religious life, and had the misfortune to be afflicted with leprosy. One of their appellations was filles meselles, in which latter word, you will immediately recognize the origin of our term for the disease still prevalent among us, the measles. Johnson strangely derives this word from morbilli; but the true northern roots have been given by Mr. Todd, in his most valuable republication of our national dictionary; a work which now deserves to be named after the editor, rather than the original compiler. It may also be added, that the word was in common use in the old Norman French, and was plainly intended to designate a slight degree of scurvy.

To pursue this subject a few steps farther, Jamieson, who is as excellent in points of etymology as Johnson is deficient, quotes, in his Scottish Dictionary, an instance where the identical expression, *meselle-houses*, is used in old English;

> "...to meselle-houses of that same rond,
>
> Thre thousand mark unto ther spense he fond."
>
> R. BRUNNE, p. 136.

The Norfolk farmers and dairy-maids tell us to this day of *measly pork*: in Scotch, a leper is called a *mesel*; and, among the Swedes, the word for measles is one nearly similar in sound, *mäss-ling*. The French academy, however, have refused to admit *meselle* to the honor of a place in their language, because it was obsolete or vulgar in the time of Louis XIIIth. The word is expressive, and no better one has supplied its place; and we may suppose that it was introduced by the Norman conquerors, and that it properly belongs to the Gothic tongues, in the whole of which the root is to be found more or less modified. Instances of this kind, and they are many, serve as additional proofs, if proofs indeed were needed, of the common origin of the Neustrian Normans, of the Lowland Scots, and of the Saxon and Belgian tribes, who peopled our eastern shores of England.

The priory continued to be appropriated to its original purpose till 1366, when Charles Vth united it to the hospital, called the Magdalen, at Rouen, upon condition that a mass should be celebrated there daily for the repose of his soul. In the year 1600, on the destruction of the abbey upon Mont Ste. Catherine, the monks of that establishment were allowed to fix themselves at St. Julien; but they resigned it, after a period of sixty-seven years, to the Carthusians of Gaillon, who, incorporating themselves with their brethren of the same order at Rouen, formed a very opulent community. The monastery, previously occupied by the latter, was known by the poetical appellation of la Rose de Notre Dame: indeed, it is thus termed in the charter of its foundation, dated 1384. But the situation was unhealthy, and the new comers had therefore little difficulty in persuading its occupants to remove to the convent of St. Julien, which they inhabited conjointly till the revolution. At a very short period before that event, they had rebuilt the whole of the priory with such splendor, that it was one of the most magnificent in the neighborhood. But the edifice, which had then been scarcely raised, was soon afterwards levelled with the ground. The foundations alone attest the former extent of the buildings; and the park, now in a state of utter neglect, their original importance.

Rouen, as I have observed, is scantily ornamented with remains of *real* Norman architecture; for, even at the risk of a bull, we must deny that title to the Norman edifices of the pointed style. Its vicinity, however, furnishes a greater number of specimens, among which the churched of *Léry*, of *Pavilly*, and of *Yainville*, are all of them deserving of a visit from the diligent antiquary.

Léry is a village adjoining Pont-de-l'Arche: its church is cruciform, having in the centre a low, massy, square tower, surmounted by a modern spire. A row of plain Norman arches, intended only for ornament, runs round the tower near the base, and over them on each side is a single round-headed window. All the other windows of the building are of the same construction, and this renders it probable that the east end, in which there is also one of these windows, is really coeval with the rest of the church; though, contrary to the usual plan of the Norman churches, it is terminated by a straight wall instead of a semi-circular apsis. The west front contains a rich Norman door-way, surmounted by three windows of the same style, adjoining each other, with a triple row of the chevron-ornament above them. The interior wears the appearance of remote antiquity: the arches are without mouldings, the pillars without bases, and the capitals are destitute of all ornamental sculpture. In fact, these portions are nothing but rounded piers; and so obviously was mere solid strength the aim of the architect, that their diameter is fully equal to two-thirds of their height. A double row of pillars and arches separates the nave into three parts, of unequal width; and another arch of greater span, though equally plain, divides it from the chancel. In St. Julien, we observe a most simple exterior, accompanied by an interior of comparatively an ornamented style: here the case is exactly the reverse; but in neither instance does there appear any reason to doubt that the whole of the building is coeval. We shall be driven, therefore, to admit, that any inferences respecting the æra of architecture drawn merely from the comparative richness of the style, must be considered of little weight, and that, even in those days, a great deal depended upon the fancy of the patron or architect. Of the real time of the erection of the church at Léry, there is no certain knowledge. Topographers, however minute in other matters, seem in general to have considered it beneath their dignity to record the dates of parish-churches; though, as connected with the history of the arts, such information is exceedingly valuable. Lauglois, who has given a figure of the western front of this at Léry, refers it without any hesitation to the time of the Carlovingian dynasty. But this opinion is merely grounded on the resemblance of some of its capitals to those of the pillars in the crypt at St. Denis; the best judges doubt whether there is a single architectural line in that crypt, which can fairly be referred to the reign of Charlemagne. Hence such a proof is entitled to little attention; and On studying the style of the whole, and its conformity with the more magnificent front of St. Georges de Bocherville, it would seem most reasonable to regard them both as of nearly the same æra, the time of the Norman Conquest. We may through them be enabled to fix the date to a specimen of ancient architecture in our own country, more splendid than these, the Church of Castle Rising, whose west front is so much on the same plan, that it can scarcely have been erected at a very different period.

Pavilly has considerably more to recommend it, as the "magni nominis umbra" than either of the others; it having been the seat of an abbey founded about the year 668, and named after Saint Austreberte, who first presided over it. Here, too, we have the advantage of being able to ascertain with greater precision the date of the building, which, in the archives of the Chartreux at Rouen[70], is stated to have been constructed about the conclusion of the eleventh century. The remains of the monastery are not considerable: they consist of little more than a ruined wall, containing three circular arches, evidently very ancient from their simplicity and the style of their masonry, and some pillars with capitals differing in ornament from any others I recollect, but imitations of the Grecian, or rather attempts to improve upon it. The inside of the parish-church is more interesting than the ruins of the abbey. It is characterised, as you will observe in the annexed sketch, by massy square piers, to each side of which are attached several small clustered columns, intended merely for ornament. One of them is fluted, the work, probably, of some subsequent time; and another, on the same pier, is truncated, to afford a pedestal for the statue of a saint. The capitals are without sculpture.

The church at Yainville differs materially from either of the others: its square low central tower is of far greater base than that of Léry: the transept parts of the cross have been demolished; and, beyond the tower, to the east, is only an addition that looks more like an apsis than a choir, a small semi-circular building with a roof of a peculiarly high pitch, like those of the stone-roofed chapels in Ireland, which, I trust, I shall be able hereafter to convince you were undoubtedly of Norman origin. But the most curious feature in this building is, that one of the buttresses is pierced with a narrow lancet window; a decisive proof, that the Normans regarded their buttresses as constituent parts of the edifice at its original construction, and that they did not add them at a subsequent time, or design them to afford support, in the event of any unexpected failure of strength. Indeed, what are usually called Norman buttresses, such as we find at Yainville, and at the lazar-house at St. Julien, have so very small a projection, that they seem much more designed to add ornament or variety than for any useful purpose.—Yainville is a parish adjoining Jumieges, and was formerly dependent upon the celebrated abbey there, which will furnish ample materials for a future letter.

Footnotes:

[63] *Taillepied, Antiquités de Rouen, p. 77.*

[64] *Vol. II. part V. p. 8.*

[65] *Seroux d'Agincourt, Historie de la Décadence de l'Art; plate 10, Sculpture, fig. 4-7.*

[66] *Du Moulin, Histoire Générale de Normandie, p. 236.*

[67] *Duchesne, Scriptores Normanni, p. 558.*

[68] *Histoire de l'Abbaye de St. Ouen, p. 188.*

[69] *Farin, Histoire de Rouen, V. p. 121*

[70] *Description de la Haute Normandie, II. p. 268.*

LETTER X
EARLY POINTED ARCHITECTURE—
CATHEDRAL—EPISCOPAL PALACE

(Rouen, June, 1818.)

In passing from the true Norman architecture, characterised "by the circular arch, round-headed doors and windows, massive pillars with a kind of regular base and capital, and thick walls without any very prominent buttresses",[71] to those edifices which display the pointed style, I shall enter into a more extensive field, and one where the difficulty no longer lies in discovering, but in selecting objects for observation and description.

The style which an ingenious author of our own country has designated as early English[72], is by no means uncommon in Normandy. In both countries, the circular style became modified into Gothic, by the same gradations; though, in Normandy, each gradation took place at an earlier period than amongst us. The style in question forms the connecting link between edifices of the highest antiquity, and those of the richest pointed architecture; combined in some instances principally with the peculiarities of the former, in others with the character of the latter: generally speaking, it assimilates itself to both. The simplicity of the principal lines betray its analogy to its predecessors; whilst the form of the arch equally displays the approach of greater beauty and perfection.

Of this æra, the cathedral[73] of Rouen is unquestionably the most interesting building; and it is so spacious, so grand, so noble, so elegant, so rich, and so varied, that, as the Italians say of Raphael, "ammirar non si può che non s'onori."—By an exordium like this, I am aware that an expectation will be raised, which it will be difficult for the powers of description to gratify; but I have still felt that it was due to the edifice, to speak of it as I am sure it deserves, and rather to subject myself to the charge of want of ability in describing, than of want of feeling in the appreciation of excellence.

The west front opens upon a spacious parvis, to which it exposes a width of one hundred and seventy feet, consisting of a centre, flanked by two towers of very dissimilar form and architecture, though of nearly equal height. Between these is seen the spire, which rises from the intersection of the cross, and which, from this point of view, appears to pierce the clouds; and these masses so combine themselves together, that the entire edifice assumes a pyramidical outline. The French, who, without any real affection for ancient architecture, are often extravagant in their praises, regard this spire as a "chef d'œuvre de hardiesse, d'élégance, et de légèreté." Bold and light it certainly is; but we must pause before we consider it as elegant: the lower part is a combination of very clumsy Roman pediments and columns; and, as it is constructed of wood, the material conveys an idea of poverty and comparative meanness.—It is commonly said in France, that the portal of Rheims, joined to the nave of Amiens, the choir of Beauvais, and the tower of Chartres, would make a perfect church; nor is it to be denied that each of these several cathedrals surpasses Rouen in its peculiar excellence; but each is also defective in other respects; so that Rouen, considered as a whole, is perhaps equal, if not superior, to any. The front is singularly impressive: it is characterised by airy magnificence. Open screens of the most elegant tracery, and filled, like the pannels to which they correspond, with imagery, range along the summit. The blue sky shines through the stone filagree, which appears to be interwoven like a slender web; but, when you ascend the roof, you find that it is composed of massy limbs of stone, of which the edge alone is seen by the observer below. This free tracery is peculiar to the pointed architecture of the continent; and I cannot recollect any English building which possesses it. The basement story is occupied by three wide door-ways, deep in retiring mouldings and pillars, and filled with figures of saints and martyrs, "tier behind tier, in endless perspective." The central portal, by far the largest, projects like a porch beyond the others, and is surmounted by a gorgeous pyramidal canopy of open stone-work, in whose centre is a great dial, the top of which partly conceals the rose window behind. This portal, together with the niches above on either side, all equally crowded with bishops, apostles, and saints, was erected at the expence of the cardinal, Georges d'Amboise, by whom the first stone was laid, in 1509[74].

The lateral door-ways are of a different style of architecture, and, though obtusely pointed, are supposed to be of the eleventh century: a plain and almost Roman circular arch surmounts the southern one. Over each of the entrances is a curious bas-relief: in the centre is displayed the genealogical tree of Christ; the southern contains the Virgin Mary surrounded by a number of saints; the northern one, the most remarkable[75] of all, affords a representation of the feast given by Herod, which ended in the martyrdom of the Baptist. Salomè, daughter of Herodias, plays, as she ought to do, the principal character. The group is of good sculpture, and curiously illustrative of the costumes and manners of the times. Salomè is seen dancing in an attitude, which perchance was often assumed by the tombesteres of the elder day; and her position affords a graphical comment upon the Anglo-Saxon version of the text, in which it is said that she "tumbled", before King Herod. The bands or pilasters (if we may so call them) which ornament the jambs of the door-ways, are crowned with graceful foliage in a very pure style; and the pedestals of the lateral pillars are boldly underworked.

On the northern side of the cathedral is situated the cloister-court. Only a few arches of the cloister now remain; and it appears, at least on the eastern side, to have consisted of a double aisle. Here we view the most ancient portion of the tower of Saint Romain.—There is a peculiarity in the position of the towers of this cathedral, which I have not observed elsewhere. They flank the body of the church, so as to leave three sides free; and hence the spread taken by the front of the edifice, when the breadth of the towers is added to the breadth of the nave and aisles. The circular windows of the tower which look in the court, are perhaps to be referred to the eleventh century; and a smaller tower affixed against the south side, containing a stair-case and covered by a lofty pyramidical stone roof, composed of flags cut in the shape of shingles, may also be of the same æra. The others, of the more ancient windows, are in the early pointed style; and the portion from the gallery upwards is comparatively modern; having been added in 1477. The roof, I suppose, is of the sixteenth century.

The southern tower is a fine specimen of the pointed architecture in its greatest state of luxuriant perfection, enriched on every side with pinnacles and statues. It terminates in a beautiful octagonal crown of open stone-work.—Legendary tales are connected with both the towers: the oldest borrows its name from St. Romain, by whom chroniclers tell us that it was built; the other is called the Tour de Beurre, from a tradition, that the chief part of the money required for its erection was derived from offerings given by the pious or the dainty, as the purchase for an indulgence granted by Pope Innocent VIIIth, who, for a reasonable consideration, allowed the contributors to feed upon butter and milk during Lent, instead of confining themselves, as before, to oil and lard.—The archbishop, Georges d'Amboise, consecrated this tower, of which the foundation was laid in 1485; and he had the satisfaction of living to see it finished, in 1507, after twenty-two years had been employed in the building.

The cardinal was so truly delighted by the beauty of the structure, which had arisen under his auspices, that he determined to grace it with the largest bell in France; and such was afterwards cast at his expence.—Even Tom of Lincoln could scarcely compete with Georges d'Amboise; for thus the bell was duly christened. It weighed thirty-three thousand pounds; its diameter at the base was thirty feet; its height was ten feet; and thirty stout and sweating bell-ringers could hardly put it into swing.—Such was the importance attached to the undertaking, that it was thought worthy of a religious ceremony. At the appointed hour for casting the bell, the clergy paraded in full procession round the church, to implore the blessing of heaven upon the work; and, when the signal was given that the glowing metal had filled the enormous mould, Te Deum resounded as with one voice; the organ pealed, the trombones and clarions sounded, and all the other bells in the cathedral joined, as loudly and as sweetly as they could, in announcing the birth of their prouder brother.—The remainder of the story is of a different complexion:—The founder, Jean le Machon, of Chartres, died from excess of joy, and was buried in the nave of the cathedral, where Pommeraye[76] tells us the tomb existed in his time; with a bell engraved upon it, and the following epitaph:—

> "Cy-dessous gist Jean le Machon
>
> De Chartres homme de façon
>
> Lequel fondit Georges d'Amboise
>
> Qui trente six mille livres poise
>
> Mil cinq cens un jour d'Aoust deuxième
>
> Puis mourut le vingt et unième."

Nor was this the only misfortune; for, after all, this great bell proved, like a great book, a great nuisance: the sound it uttered was scarcely audible; and, at last, in an attempt to render it vocal, upon a visit paid by Louis XVIth to Rouen in 1786, it was cracked[77]. It continued, however, to hang, a gaping-stock to children and strangers, till the revolution, in 1793, caused it to be returned to the furnace, whence it re-issued in the shape of cannon and medals, the latter commemorating the pristine state of the metal with the humiliating legend, "monument de vanité détruit pour l'utilité[78]."

Some of the clerestory windows on the northern side of the nave are circular: the tracery which fills them, and the mouldings which surround them, belong to the pointed style; the arches may therefore have been the production of an earlier architect. The windows of the nave are crowned by pediments, each terminating, not with a pinnacle, but with a small statue. The pediments over the windows of the choir are larger and bolder, and perforated as they rise above the parapet; the members of the mouldings are full, and produce a fine effect.

The northern transept is approached through a gloomy court, once occupied by the shops of the transcribers and caligraphists, the libraires of ancient times, and from them it has derived its name. The court is entered beneath a gate-way of beautiful and singular architecture, composed of two lofty pointed arches of equal height, crowned by a row of smaller arcades. On each side are the walls of the archiepiscopal palace, dusky and shattered, and desolate; and the vista terminates by the lofty Portal of St. Romain; for it is thus the great portal of the transept is denominated. The oaken valves are bound with ponderous hinges and bars of wrought iron, of coeval workmanship. The bars are ornamented with embossed heads, which have been hammered out of the solid metal. The statues which stood on each side of the arch-way have been demolished; but the pedestals remain. These, as well as other parts of the portal, are covered with sculptured compartments, or medallions, in high preservation, and of the most singular character. They exhibit an endless variety of fanciful monsters and animals, of every shape and form, mermaids, tritons, harpies, woodmen, satyrs, and all the fabulous zoology of ancient geography and romance; and each spandril of each quatrefoil contains a lizard, a serpent, or some other worm or reptile. They have all the oddity, all the whim, and all the horror of the pencil of Breughel. Human groups and figures are interspersed, some scriptural, historical, or legendary; others mystical and allegorical. Engravings from these medallions would form a volume of uncommon interest. Two lofty towers ornament the transept, such as are usually seen only at the western front of a cathedral. The upper story of each is perforated by a gigantic window, divided by a single mullion, or central pillar, not exceeding one foot in circumference, and nearly sixty feet in height. These windows are entirely open, and the architect never intended that they should be glazed. An extraordinary play of light and shade results from this construction. The rose window in the centre of the transept is magnificent: from within, the painted glass produces the effect of a kaleidoscope. — The pediment or gable of this transept was materially injured by a storm, in 1638, one hundred and thirty years after it was completed; and the damage was never restored.

The southern transept bears a near resemblance to that which I have already described; but it was originally richer in its ornaments, and it still preserves some of its statues. Here the medallions relate chiefly to scripture-history; but the sculpture is greatly corroded by the weather, and the more delicate parts are nearly obliterated; besides which, as well here, as at the other entrances, the Calvinists, in 1562, and, more recently, the Revolutionists, have been most mischievously destructive, mutilating and decapitating without mercy. The spirit, indeed, of the French reformers, bore a near resemblance to the proceedings of John Knox and his brethren: the people embraced the new doctrine with turbulent violence. There was in it nothing moderate, nothing gradual: it was not the regular flow of public opinion, undermining abuses, and bringing them slowly to their fall; but it was the thunderbolt, which—

"In sua templa furit, nullâque exire vetante

Materiâ, magnamque cadens magnamque revertens

Dat stragem latè sparsosque recolligit ignes."

Among the legends recorded on the southern portal, or the *Portail de la Calende*, is that of the corn-merchant; the confiscation of whose property paid, as the chronicles tell us, for the erection of this beautiful entrance. He himself, if we may believe the same authority, was hanged in the street opposite to it, in consequence of having been detected in the use of false measures.

The original Lady-Chapel, at the east end of the cathedral, was taken down in 1302. The present, which is considerably more spacious, is chiefly of a date immediately subsequent. Part, however, was built in 1430, when new and larger windows were inserted throughout the church; whilst other parts were not finished till 1538, at which time the Cardinal Georges d'Amboise restored the roof of the choir, which had been injured in 1514, by the destruction of the spire.

The square central tower, which is low and comparatively plain, is the work of the year 1200. It is itself more ancient than would be supposed from the character of its architecture; but it occupies the place of one of still greater antiquity, which was materially damaged in 1117, when the original spire of the church was struck by lightning. This first spire was of stone, but was replaced by another of wood, which, as I have just mentioned, was also destroyed at the beginning of the sixteenth century. A fire, arising from the negligence of plumbers employed to repair the lead-work, was the cause of its ruin.—To remedy the misfortune, recourse was had to extraordinary efforts: the King contributed twelve thousand francs; the chapter a portion of their revenue and their plate; collections were made throughout the kingdom; and Leo Xth authorised the sale of indulgences, a measure, which, at nearly the same period, in its more extensive adoption for the building of St. Peter's at Rome, shook the Papacy to its foundation. The spire thus raised, the second of wood, but the third in chronological order, is the one which is now in existence. It was, like its predecessor, endangered by the carelessness of the plumbers, in 1713; but it does not appear to have required any material reparations till ten years ago, when a sum of thirty thousand francs was expended upon it.

From what has already been said, you will not have failed to observe that this cathedral is the work of so many different periods, that it almost contains within itself a history of pointed architecture. To attempt a labored description of it were idle: minute details of any one of the portals would fill a moderate volume; and a quarto of seven hundred pages, from which I have borrowed most of my dates, has already been written upon the subject by a Benedictine Monk of the name of Pommeraye, who also published the history of the Archbishops of the See[79].

The first church at Rouen was built about the year 270: three hundred and thirty years subsequently, this edifice was succeeded by another, the joint work of St. Romain and St. Ouen, which was burned in the incursions of the Normans, about the year 842. Fifty years of Paganism succeeded; at the expiration of which period, Rollo embraced the faith of Christ, and Rouen saw once more within its walls, by the munificence and piety of the conqueror, a place of Christian worship. Richard Ist, grandson of this duke, and his son Robert, the archbishop, enlarged the edifice in the middle of the tenth century; but it was still not completed till 1063, when, according to Ordericus Vitalis, it was dedicated by the Archbishop Maurilius with great pomp, in the presence of William, Duke of Normandy, and the bishops of the province. Of this building, however, notwithstanding what is said by Ducarel[80] and other authors, it is certain that nothing more remains than the part of St. Romain's tower, just noticed, and possibly two of the western entrances; though the present structure is believed to occupy the same spot.

To the honor of the spirit and good feeling of the inhabitants of Rouen, this church is one of those that suffered least in the outrages of the year 1793. Its dimensions, in French feet, are as follows:—

	FEET.
Length of the interior	408
Width of ditto	83
Length of nave	210
Width of nave	27
Ditto of aisles	15
Length of choir	110
Width of ditto	35½
Ditto of transept	25½
Length of ditto	164
Ditto of Lady-Chapel	88
Width of ditto	28
Height of spire	380
Ditto of towers at the west end	230

Ditto of nave	84
Ditto of aisles and chapels	42
Ditto of interior of central tower	152
Depth of chapels	10

Four clustered pillars support the central tower, each of which is thirty-eight feet in circumference; the rest, of which there are forty-four in the nave and choir, those in the former clustered, the others circular, are less by one-third. The windows amount in number to one hundred and thirty-three; the chapels to twenty-five. Most of the latter were fitted up during the minority of Louis XIVth, with wreathed columns, entwined with foliage, the style in vogue in the seventeenth century. In the farthest of these chapels, upon the south side, is the tomb of Rollo, first Duke of Normandy; in the opposite chapel, that of his son and successor, William Longue-Epeé, who was treacherously murdered at Pecquigny, in 944, during a conference with Arnoul, Count of Flanders.

Statue of Rollo in Rouen Cathedral

The effigies of both these princes still remain placed upon sarcophagi, under plain niches in the wall. They are certainly not contemporary with the persons which they represent, but are probably productions of the thirteenth century, to which period Mr. Stothard, from whose judgment few will be disposed to appeal, refers the greater part of what are called the most ancient in the Musée des Monumens Français. At the same time, they may possibly have been copied from others of earlier date; and I therefore send you a slight sketch of the figure of Rollo. Even imaginary portraits of celebrated men are not without their value: we are interested by seeing how they have been conceived by the artist. — Above the statue is the following inscription: —

HIC POSITUS EST

ROLLO,

NORMANNIÆ A SE TERRITÆ, VASTATÆ,

RESTITUTÆ,

PRIMUS DUX, CONDITOR, PATER,

A FRANCONE ARCHIEP. ROTOM.

BAPTIZATUS ANNO DCCCCXIII,

OBIIT ANNO DCCCCXVII.

OSSA IPSIUS IN VETERI SANCTUARIO,

NUNC CAPITE NAVIS, PRIMUM CONDITA,

TRANSLATO ALTARI, HIC COLLOCATA

SUNT A B. MAURILIO ARCHIEP. ROTOM.

ANNO MLXIII.

Two other epitaphs in rhyming Latin, which were previously upon his tomb, are recorded by various authors: the first of them began with the three following lines —

DUX NORMANNORUM, CUNCTORUM NORMA BONORUM,

ROLLO FERUS FORTIS, QUEM GENS NORMANNICA MORTIS

INVOCAT ARTICULO, CLAUDITUR HOC TUMULO.

Over William Longue-Epeé is inscribed —

HIC POSITUS EST

GULIELMUS DICTUS LONGA SPATHA,

ROLLONIS FILIUS,

DUX NORMANNIÆ,

PREDATORIE OCCISUS DCCCCXXXXIV.

with an account of the removal of his bones, exactly similar to the concluding part of his father's epitaph.

The perspective on first entering the church is very striking: the eye ranges without interruption, through a vista of lofty pillars and pointed arches, to the splendid altar in the Lady-Chapel, which forms at once an admirable termination to the building and the prospect. The high altar in the choir is plain and insulated. No other praise can be given to the screen, except that it does not interrupt the view; for surely it was the very consummation of bad taste to place in such an edifice, a double row of eight modern Ionic pillars, in white marble, with the figures of Hope and Charity between them, surmounted by a crucifix, flanked on either side with two Grecian vases.

The interior falls upon the eye with boldness and regularity, pleasing from its proportions, and imposing from its magnitude. The arches which spring from the pillars of the aisles, are surmounted by a second row, occupying the space which is usually held by the triforium: the vaulted roof of the aisles runs to the level of the top of this upper tier. This arrangement, which is found in other Norman churches, is almost peculiar to these; and in England it has no parallel, except in the nave of Waltham Abbey. Within the aisle you observe a singular combination of small pillars, attached to the columns of the nave: they stand on a species of bracket, which is supported by the abacus of the capital; and they spread along the spandrils of the arches on either side. These pillars support a kind of entablature, which takes a triangular plan. The whole bears a near resemblance to the style of the Byzantine architecture. Above the second row of arches are two rows of galleries. The story containing the clerestory windows crowns the whole; so that there are five horizontal divisions in the nave.—I give these details, because they indicate the decided difference of order which exists between the Norman and the English Gothic; a difference for which I have not been able to assign any satisfactory cause.

The tombs that were originally in the choir, commemorating Charles Vth, of France; Richard Cœur de Lion; his elder brother, Henry; and William, son of Geoffrey Plantagenet, were all removed in 1736, as interfering with the embellishments then in contemplation. The first of them alone was preserved and transferred to the Lady-Chapel, where it has subsequently fallen a victim to the revolution. The others are wholly destroyed; nor could Ducarel find even a fragment of the effigies that had been upon them; but engravings of these had fortunately been preserved by Montfaucon[81], from whom he has copied them. The monument of the celebrated John of Lancaster, third son of our Henry IVth, better known as the Regent Duke of Bedford, had been previously annihilated by the Calvinists. Lozenge-shaped slabs of white marble, charged with inscriptions, were inserted in the pavement over the spots that contain the remains of the princes, and they have been suffered to continue uninjured through the succeeding tumults. On the right of the altar, you read,—

COR
RICHARDI, REGIS ANGLIÆ,
NORMANNIÆ DUCIS,
COR LEONIS DICTI.
OBIIT ANNO
MCXCIX.

On the opposite side:—

HIC JACET

HENRICUS JUNIOR,
RICHARDI, REGIS ANGLIÆ,
COR LEONIS DICTI, FRATER.
OBIIT ANNO

MCLXXXIII.

And in the choir behind the altar:—

AD DEXTRUM ALTARIS LATUS

JACET

JOHANNES, DUX BEDFORDI,

NORMANNIÆ PROREX.

OBIIT ANNO

MCCCCXXXV.

Of Prince William nothing is said; it was found, upon opening his place of sepulture, that he had not been interred here.—Richard strangely received a triple funeral. In obedience to his wishes, his heart was buried at Rouen, while his body was carried to Fontevraud, and his entrails were deposited in the church of Chaluz, where he was killed:—this division is commemorated in the quaint, yet energetic lines, which are said to have been inscribed upon his tomb:—

VISCERA CARCEOLUM, CORPUS FONS SERVAT EBRARDI,

ET COR ROTOMAGUM, MAGNE RICHARDE, TUUM.

IN TRIA DIVIDITUR UNUS QUI PLUS FUIT UNO;

NEC SUPEREST UNI GLORIA TANTA VIRO.

Richard neither withheld his gifts nor his protection from the metropolitan church; and, after his death, the chapter inclosed the heart of their benefactor in a shrine of silver. But a hundred and fifty years subsequently, the shrine was despoiled, and the precious metal was melted into ingots, forming a portion of the ransom which redeemed St. Louis from the fetters of his Saracen conqueror.

Henry the younger, who was crowned King of England during the life-time of his father, against whom he subsequently revolted, also requested on his death-bed, that his body might be interred in this church; and his directions were obeyed, though not without much difficulty; for the chapter of the cathedral of Mans, where his servants rested with the body *in transitu*, seized and buried it there; nor did those of Rouen recover the corpse, without application to the Pope and to the King his father.

A tablet of black marble, affixed to one of the pillars of the nave, contains the following interesting memorial:

IN MEDIA NAVI,

E REGIONE HUJUS COLUMNÆ,

JACET

BEATÆ MEM. MAURILIUS,

ARCHIEP. ROTOM. AN. MLV.

HANC BASILICAM PERFECIT

CONSECRAVITQUE ANNO MLXIII.

VIX NATOS BERENGARII ERRORES

IN PROX. CONCIL. PRÆFOCAVIT.

PLENUS MERITIS OBIIT ANN. MLXVII.

HOC PONTIF. NORMANNI,

GULIELMO DUCE, ANGLIA POTITI SUNT

ANNO MLXVI.

In the northern aisle of the choir, there still exists a curious monument, in an injured state indeed, but well deserving of attention, from its antiquity. It has been referred by tradition to Maurice, or William of Durefort, both of them archbishops of Rouen, and buried in the cathedral, the former in 1237, the latter in 1331; but the recumbent figure upon it seems of a yet more distant date. It differs in several respects from any that I have seen in England[82]. The tomb is in the wall, behind a range of pillars, which form a kind of open screen round the apsis. Below the effigy, it is decorated with a row of whole-length figures of saints, much mutilated: the circular part above is lined with angels, a couple of whom are employed in conveying the soul of the deceased in a winding-sheet to heaven[83].

Tomb of an Archbishop, Rouen Cathedral.

The Lady-Chapel contains two monuments of great merit, and which, considered as specimens of matured art, have now no rivals in Normandy; for both owe their origin to a period of refinement and splendor. The sepulchre raised over the bodies of the two Cardinals of Amboise, successively Archbishops of Rouen, towers on the southern side of the chapel. The statues of the cardinals are of white marble. The prelates appear kneeling in prayer; and the following inscription, engraved in a single line, and not divided into verses, is placed beneath them:—

PASTOR ERAM CLERI, POPULI PATER, AUREA SESE

LILIA SUBDEBANT QUERCUS[84] ET IPSA MIHI.

MORTUUS EN JACEO, MORTE EXTINGUUNTUR HONORES;

AT VIRTUS MORTIS NESGIA MORTE VIRET.

Immediately behind the cardinals are figures of patron saints; a centre tablet represents St. George and the Dragon; above are the apostles; below, the seven cardinal virtues. The execution of these is particularly admired, especially that of the figure of Prudence; but a row of still smaller figures, in devotional attitudes, carved upon the pilasters between the virtues, are in higher taste. Various arabesques in basso-relievo, of great beauty, and completely in the style of the Loggie of Raphael, adorn the other parts of this sumptuous tomb.—As a whole it is unquestionably grand, and it is yet farther valuable as an illustration of the gorgeous taste that prevailed at the end of the fifteenth century; but the mixture of black and white marble and gilding has by no means a good effect, and every part is overloaded with ornaments[85]. These, however, are the faults of the times: its merits are its own.

On the north side of the chapel is entombed the Duke of Brezé, once Grand Seneschal of Normandy; his tomb is chaste and simple, forming a pleasing contrast to the elaborate memorial of the cardinals. The statue of the seneschal himself, represented stretched as a corpse, upon a black marble sarcophagus, is admirable for its execution. The rigid expression of death is visible, not only in the countenance, but extends through every limb. Diana of Poitiers, a beauty who enjoys more celebrity than good fame, erected the monument; and she caused her statue to be placed on the tomb, where she is seen kneeling and contemplating. In the following inscription she promises to be as faithful and united to him after his death as she was while they both lived: and she truly kept her word; for, during his life-time, she was grievously suspected of infidelity[86], and she subsequently lived in an open state of concubinage with Henry IInd, and was at last buried at her own celebrated residence at Anet, twenty leagues from her husband.—

> *HOC, LODOICE, TIBI POSUI, BREZÆE, SEPULCHRUM,*
>
> *PICTONIS AMISSO MOESTA DIANA VIRO;*
>
> *INDIVULSA TIBI QUONDAM ET FIDISSIMA CONJUX,*
>
> *UT FUIT IN THALAMO, SIC ERIT IN TUMULO.*

A second female figure on the tomb, with a child in her arms, has been supposed intended to represent the nurse of the duke; as if the design of the sculptor had been to read a lesson to mortality, by exhibiting the warrior in the helplessness of infancy, in the vigor of manhood, and as a breathless corpse. Some persons, however, consider it as a personification of Charity; others suppose that it represents the Virgin Mary. In the midst was originally an erect statue of De Brezé, decorated with the various symbols of his dignities; but this sinned beyond the hope of redemption against the doctrines of liberty and equality, and it was accordingly removed at the time of the revolution, together with two inscriptions. One of them, which detailed his honors, with the addition that he died July twenty-third, 1531, has recently been recovered by the care of M. Riaux, and is restored to its place. The other inscription and the effigy, it is feared, are irrevocably lost. An equestrian statue in the upper part of the monument was suffered to remain, and, as a record of the military costume of the sixteenth century, I annex a sketch of it. The armorial hearings upon the horse and armor are nearly obliterated.—The pile is surmounted a figure of Temperance; the bridle in whose mouth shews how absurd is allegory, when "submitted to the faithful eye."

Lenoir, who, in his work on the Musée des Monumens Français, has treated much at large of the history of Diana of Poitiers, and has figured her own beautiful mausoleum, which he had the merit of rescuing from destruction, pronounces[87] this monument to be from the hand of Jean Cousin, one of the most able sculptors of the French school.

Over the altar in the Lady-Chapel is the only good painting in the cathedral, the *Adoration of the Shepherds*, by Philip de Champagne, a solid, well-colored, and well-grouped picture. Two cherubs in the air are excellently conceived and drawn: the whole is lighted from the infant Christ in the cradle, a *concetto*, which has been almost universally adopted, since the time when Corregio painted his celebrated *Notte*, now at Dresden.

There is no great quantity of painted glass in the church, but much of it is of good quality. The windows of the choir, on either side of the Lady-Chapel, are as rich as a profusion of brilliant colors can make them; but the figures are so small, and so crowded, that the subjects cannot be traced. They are said to be the work of the thirteenth century. The painted windows in St. Stephen's chapel, of the sixteenth century, are generally considered the best in the cathedral. I own, however, that I should give the preference to those in the chapel of St. Romain, in the south transept. One of them is filled with allegorical representations of the virtues of the archbishop; another with his miracles: every part is distinct and clear, and executed with great force and great minuteness. The vestments of the saint have all the delicacy of miniature-painting.

The library of the cathedral, formerly one of the richest in France, disappeared during the revolution; but the noble room which contained it, one hundred feet long, by twenty-five feet wide, still remains uninjured; as does the door which led into it from the northern transept, and which continues to this day to bear the inscription, *Bibliotheca*. The staircase, communicating with this door, is delicate and beautiful. The balustrades are of the most elegant filagree; and it has all the boldness and lightness which peculiarly characterise the French Gothic. Its date being well ascertained, we may note it as an architectural standard. It was erected by the archbishop, Cardinal d'Etouteville, about the year 1460, thirty or forty years subsequently to the building of the room.

Respecting the contents of the sacristy, I can say little from my own knowledge; but I find by Pommeraye, that, before the revolution, it boasted of a large silver image of the Virgin, endued with peculiar sanctity, a few drops of her milk, and a portion of her hair[88]; a splinter of the true cross, set in gold, studded with pearls, sapphires, and turquoises; and reliques of saints without number. Now, however, it appears, that of all its treasures, it has preserved little else except the shrine of St. Romain, and another known by the general name of Chasse des Saints. The former is two feet six inches long, and one foot nine inches high, and is of handsome workmanship, with a variety of figures on the sides, and St. Romain himself at the top. Formerly it was supposed to be made of gold; now I was assured by one of the canons, that it is of silver gilt; but Gilbert[89], who is a plain layman, maintains that it is only copper. Had it been otherwise, it would have contributed to the ways and means of the unchristian republic; but the democrats spared it, for they had well ascertained that the metal was base, and that the jewels, which adorn it, are but glass.—This is not the original shrine which held the precious relics: the shrine in which they were deposited by the archbishop, William Bonne Ame, when first brought to the cathedral, in 1090, was sold during a famine, and its proceeds distributed to the starving poor; after which, in 1179, Archbishop Rotrou caused another still more costly to be made; but the latter was broken to pieces by the Calvinists, in 1562, and the saint's body cast into the fire[90].

Thus, then, I have led you, as far as I am able; through the cathedral, adjoining which, at the east end, stands the palace of the archbishop, a large building, but neither handsome nor conspicuous, principally the work of the Cardinal Georges d'Amboise, though begun by the Cardinal d'Etouteville, in 1461. The rooms in it which are shewn to strangers are the anti-chamber, commonly called la salle de la Croix, the library, and the great gallery. This last, which is one hundred and sixty feet long, is also known by the name of la salle des Etats. In it are placed four very large paintings by Robert, an eminent French artist of comparatively modern date. They represent the city of Rouen, the town of Dieppe, that of Havre de Grace, and the archiepiscopal palace at Gaillon. The view of Rouen represents in the foreground the petit Château, and is on that account peculiarly interesting. All of them are fine paintings, but much injured by the damp. In the anti-chamber are portraits of seven prelates of the see, and among them those of the Cardinal de la Rochefoucault, and M. de Tressan: our guide could name no others.

The present archbishop is the Cardinal Cambacérés, brother to the ex-consul of that name, a man of moral life and regular in his religious duties. He was placed here by Napoléon, all of whose appointments of this nature, with one or two exceptions, have been suffered to remain; but I need scarcely add that, though the title of archbishop is left, and its present possessor is decorated with the Roman purple, neither the revenue, nor the dignity, nor the establishment, resemble those of former times. The chapter, which, before the revolution, consisted of an archbishop, a dean, fifty canons, and ten prebendaries, besides numberless attendants, now consists but of his eminence, with the dean, the treasurer, the archdeacon, and twelve canons. The independent annual income of the church, previous to the revolution, exceeded one hundred thousand pounds sterling; but now its ministers are all salaried by government, whose stated allowance, as I am credibly informed, is to every archbishop six hundred and twenty-five pounds per annum; to every bishop four hundred and sixteen pounds thirteen shillings and fourpence; and to every canon forty-one pounds thirteen shillings and four-pence. But each of these stipends is doubled by an allowance of the same amount from the department; and care is taken to select men of independent property for the highest dignities.—From the foregoing scale, you may judge of the state of the religious establishment in France. It is, indeed, unjustly and unreasonably depressed, and there is much room for amendment; but we must still hope and trust that things will not soon regain their former standard, though attempts are daily making to identify the

Catholic clergy with the present dynasty; and the most lively expectations are entertained from the well-known character of some of the royal family.

Footnotes:

[71] Bentham, History of Ely, 2nd edit. I. p. 34.

[72] Liverpool Panorama of Arts and Sciences, article Architecture.

[73] The only views of the cathedral with which I am acquainted, are,

A single plate of the west front, 16 in. by 11-1/2in.—Anonymous;

. . . . north side, 16 in. by 11-1/2in.—Marked S.L.B.;

A small north-west view, engraved by Pouncey, in the first volume of Gough's Alien Priories;

And the west front, on an extremely reduced; scale, in Seroux

d'Agincourt's Histoire de l'Art par les Monumens, Architecture, t. 64. f. 21. p. 68.

[74] This great benefactor to Rouen died the following year, deeply lamented by the inhabitants, and generally so by France; but, above all, regretted by Louis XIIth, his sovereign, whom, to use the words of Guicciardini, he served as oracle and authority. The author of the History of the Chevalier Bayard, is still louder in his praise.—The western facade of the cathedral was not finished till 1530, twenty years after his death.

[75] A representation of this has recently been published from an engraving on stone by Langlois.

[76] Histoire de l'Eglise Cathédrale de Rouen, p. 50.

[77] Noel, Essais sur le Département de la Seine Inférieure, II. p. 239.

[78] Millin, Histoire Métallique de la Révolution Française, t. 22. f. 84.

[79] Histoire des Archevêques de Rouen, folio 1667.

[80] Anglo-Norman Antiquities, p. 12.

[81] Monumens de la Monarchie Française, II. t. 15. f. 3 and 5.

[82] As these effigies are in general little understood, even by those who look at them with pleasure as specimens of art, or with respect as relics of antiquity, I am happy to be able to give the following detailed illustration of this at Rouen, extracted from a letter which the Right Rev. Dr. Milner had lately the kindness to write me upon the subject.

"The sepulchral monument in the cathedral of Rouen represents a prelate; that is to say, Bishop or Mitred Abbot, as appears by his mitre, gloves, ring, and sandals. But, as he bears the Pallium, (to be seen on his neck, just above his breast, and hanging down before him, almost to his feet) it appears that he is a Metropolitan, or Archbishop, as, indeed, each of the bishops of Rouen was, from the time of St. Ouen and St. Romanus, in the seventh century, if not from that of St. Nicasius, in the third or fourth. The statue has been mutilated in the mitre, the face, and the crosier; probably when the Huguenots were masters of the city. The mitre is low, as they used to be from the tenth century, when they began to rise at all in the Latin Church, down to the fourteenth, since which they have grown to their present disproportioned height. The arms are crossed, as in prayer; and the left arm supported a crosier, the remnant of which is seen under that arm. Both hands are wrapped up in ornamented gloves, which were an essential part of the prelatic dress. The principal vestment is the Planeta, Casula, or Chausible; as it was shaped till within these three or four hundred years. Underneath that, and behind the hanging Pallium, appears the Dalmatic, edged with gold lace; and under that, extending the whole breadth of the figure, and finishing with rich and deep thread lace, is the Alb, made of fine linen. The Tunic is quite hidden by the dalmatic. The Sandals appear to be of gold tissue, and to rest on a rich carpet.

"I ought to have mentioned, that the mitre appears, by the jewels with which it is ornamented, to represent that which is called Mitra pretiosa, from this circumstance. An inferior kind of mitre, worn on less solemn occasions, was termed Mitra Aurifrygiata; and a common one, made of plain linen or silk, was termed Simplex Mitra. The only part of the dress which puzzles me, is the great ornament on the neck and shoulders. The question is, (which those can best determine who have seen the original statue,) whether it adheres to the Pallium, or to the Casula. In either case, it must be considered as part of the vestment to which it adheres.

"It is quite out of my power to determine, or even to conjecture on any rational grounds, which, of a certain three-score of archbishops of Rouen, the figure represents; but, if I were to choose between Maurice, the fifty-fourth archbishop, who died in 1235, and William, of Durefort, the sixty-first, who died in 1330, from the comparative lowness of the mitre, and some other circumstances of the dress, I should determine in favor of the former. Perhaps it may represent our Walter, who was first Bishop of Lincoln, and then transferred to Rouen, by Pope Lucius IIIrd. He died in 1208, after having signalized himself as much as any of his predecessors or successors have done.

"P.S. On consulting with an intelligent ecclesiastic of Rouen, I am inclined to think that the above-mentioned ornament upon the shoulders, is the Mozetta, being a short round cloak, which all bishops still wear, with the Rochet, Pectoral Cross, and Purple Cassock, as their ordinary dress; but, in modern times, the Mozetta is laid aside, when the prelate puts on his officiating vestments; though he retains the cassock, cross, and rochet, underneath them. My informant says, that this mozett is common on the tombs of bishops who died in former ages."

[83] The same idea is to be observed on many ancient monuments: among others, it is engraved on the fine sepulchral brass to the memory of Sir Hugh Hastings, in Elsing church. —See Cotman's Norfolk Sepulchral Brasses.

[84] By the words Lilia and Quercus, are designated the armorial bearings of the King of France, and Pope Julius IInd, of the House of Rovere.

[85] The bodies of the Cardinals d'Amboise were dug up in 1793, together with most of the others interred in the cathedral, for the sake of their leaden coffins: at the same time the lead was also stripped from the transepts; and a

colossal statue of St. George, which stood on the eastern point of the choir, was likewise consigned to the furnace.

[86] Ducarel says (Anglo-Norman Antiquities, p. 20.) that she was the favorite mistress of two successive kings; but I do not find this assertion borne out by history.

[87] Vol. IV. p. 47.

[88] The doctrine of the Assumption of the Virgin Mary, gave rise to some curious doubts respecting the authenticity of the Virgin's hair. Ferrand, the Jesuit, states the arguments to the contrary with candor; but replies to them with laudable firmness. The passage is a whimsical specimen of the style and reasoning of the schools:—"Restat posteriore loco de capillis Deiparæ Virginis paucis dicere, enimverò an illi sint jam in terris!--Dubitationem aliquam afferre potest mirabilis ipsius anastasis, et in coelum viventis videntisque assumptio triumphalis.—Quid ita?—quid si intra triduum ad vitam revocata, si coelis triumphantis in morem invecta, si corpore gloriâ circumfuso Christo assidet? Quidquid Virgineo capiti crinium inerat hand dubiè cælis intulit, ne quid perfectæ ac numeris omnibus absolutæ ipsius pulchritudini deesse possit. Næ ille in politiori literaturâ imo et in rebus humanis omnino peregrinus sit qui ignoret quantum ad muliebrem formam comæ conferat pulchritudo ... ne singulas Marianæ pulchritudinis dotes persequar, ejus ima cræaries de quâ, agimus tantæ fuit venustatis ut mysticus ipsius Sponsus blandè querulus exclamare cogatur, vulnerasti cor meum in uno crine colli tui.... Nænias igitur occinere videtur qui Deiparæ capillos in terris relatos esse memoret atque adeo servari obfirmatè asseveret, cùm illos tantum ad redivivæ Virginis speciem conferre constet.—Non efficiet tamen unquam hæc Antidicomarianitæ fabula, quin credam bene multos ex aureâ Dei Genitricis cæsarie crines, diversis in locis ecclesiisque religiosè servari.... Meæ fidei non unum est argumentum; nam a primâ ætate ad confectam usque, e Marianâ comâ non pancos, ut fit, capillos pecten decussit, nisi si fortè cæsariem B. Virginis impexam semper perstitisse velis, quòd numquam (ut inquit de Christo Diva Brigitta) super eam venit vermis, aut perplexitas, aut immunditium. At sine causâ multiplicari miracula quis æquo animo feret?—Ubi vero Genetrix e vitâ discessit, quàm sollicitè pollinctrices auream illam Marianæ comæ

segetem demessuerunt, quàm in sacris suis tunc hierothecia reconderent ad memoriam tantæ Imperatricis, et ad suæ consolationis et pietatis argumentum: quòd si fortè totam funditùsque a pollinctricibus, Deiparæ reverentissimis, demessam cæsariem ferre nec possis nec velis, extremes saltem illius cincinnos attonsos fuisse feres ab piissimis illis fæminis, quibus vel perexiguus Dei Genitricis capillus ingentis thesauri loco futurus etat."—Disquisitio Reliquiaria, l. 1. cap. II.

[89] Description Historique de l'Eglise de Notre Dame de Rouen, p. 83.

[90] The event is described in the metrical history of Rouen, composed by a minstrel ycleped Poirier, the limper. This little tract is a chap-book at Rouen: most towns, in the north of France and Belgium, possess such chronicle ballads in doggerel rhyme, which are much read, and eke chaunted, by the common people.

"... un massacre horrible

Survint soudainement.

Les Huguenots terribles

Et Montgommerie puissant,

Par cruels enterprises

Renverserent les Eglises

De Rouen pour certain.

Sans aucune relâche

Pillent et volent la châsse

Du corps de St. Romain.

"Le zelé Catholique

Poursuivant l'Huguenot

Un combat héroique

Lui livra à propos,

Au lieu nommé la Crosse,

Et reprirent par force

La châsse du Patron.

Puis de la Rue des Carmes

La portent à Notre Dame

En déposititon!"

LETTER XI
POINTED ECCLESIASTICAL ARCHITECTURE—
THE CHURCHES OF ST. OUEN, ST. MACLOU,
ST. PATRICE, AND ST. GODARD

(Rouen, June, 1818.)

In the religious buildings, the subject of my preceding letters, I have endeavored to point out to you the specimens which exist at Rouen, of the two earliest styles of architecture. The churches which I shall next notice belong to the third, or *decorated* style, the æra of large windows with pointed arches divided by mullions, with tracery in flowing lines and geometrical curves, and with an abundance of rich and delicate carving.

This style was principally confined in England to a period of about seventy years, during the reigns of the second and third Edward. In France it appears to have prevailed much longer. It probably began there full fifty years sooner than with us, and it continued till it was superseded by the revival of Grecian or Italian architecture. I speak of France in general, but I must again repeat, that my observations are chiefly restricted to the northern provinces, the little knowledge which I possess of the rest being derived from engravings. No where, however, have I been able to trace among our Gallic neighbors the existence of the simple *perpendicular* style, which is the most frequent by far in our own country, nor of that more gorgeous variety denominated by our antiquaries after the family of Tudor.

So long as Normandy and England were ruled by the same sovereign, the continual intercourse created by this union caused a similarity in their architecture, as in other arts and customs; and therefore the two earliest styles of architecture run parallel in the two countries, each furnishing the counterpart of the other. Whether or not the decorated style was transmitted to England from the continent, is a question which cannot be solved, until our collections of continental architecture shall become more extensive. After the reign of Henry VIth, our intercourse with Normandy wholly ceased; and, left to ourselves, many innovations were gradually introduced, which were not known to the French architects, who, with nicer taste, adhered to the pure style which we rejected. Hence arose the perpendicular style of pointed architecture, a style sufficiently designated by its name, and obviously distinguished from its predecessors, by having the mullions of its windows, its ornamental pannelling, and other architectural members and features, disposed in perpendicular lines. Finally, however, both countries discarded the Gothic style, though at different æras. The revival of the arts in Europe, in consequence of the capture of Constantinople and of the greater commercial intercourse between transalpine Europe and Italy, gradually gave rise to an admiration of the antique: imitation naturally succeeded admiration; and buildings formed upon the classical model generally replaced the Gothic. Italian architects found earlier patrons and earlier scholars, in France, than amongst us, our intermediate style being chiefly distinguished by its clumsiness.

I will not detain you by any attempt at a comparison between the relative beauties of the Gothic and Grecian architecture, or their respective fitness for ecclesiastical buildings. The very name of the former seems sufficient to stamp its inferiority; and perhaps you will blame the employment of a term which was obviously intended at the outset as an expression of contempt; but I still retain the epithet, as one generally received, and therefore, commonly understood. It may be added, that the modern French seem to be the only Goths, in the real and true acceptation of the word. They, to the present day, build Gothic churches; but, instead of confining themselves to the prototypes left them, they are eternally aiming at alterations, under the specious name of improvements. Horace was indignant that, in the Augustan age, the meed of praise was bestowed only upon what was ancient: the architects of this nation of recent date seem under the influence of an opposite apprehension. They build upon their favorite poet:—

"Loin d'ici ce discours vulgaire

Que l'art pour jamais dégénère,

Que tout s'éclipse, tout finit;

La nature est inépuisable,

Et le génie infatigable

Est le Dieu qui la rajeunit."

But they overlook, what Voltaire makes an indispensable requisite, that art must be under the guidance of genius: when it is not so, and caprice holds the reins, the result cannot fail to be that medley of Grecian, Norman, Gothic, and Gallic, of which this country furnishes too many examples.

The church of St. Ouen is unquestionably the noblest edifice in the pointed style in this city, or perhaps in France; the French, blind as they usually are to the beauties of Gothic architecture, have always acknowledged its merits. Hence it escaped the general destruction which fell upon the conventual churches of Rouen, at the time of the revolution; though, during the violence of the storm, it was despoiled and desecrated. At one period, it was employed as a manufactory, in which forges were placed for making arms; at another, as a magazine for forage.

Nor was this the first instance of its being violated; for, like most of the religious buildings at Rouen, it was visited in the sixteenth century with the fury of the Calvinists[91], who burned the bodies of St. Ouen, St. Nicaise, and St. Remi, in the midst of the temple itself; and cast their ashes to the winds of heaven. The other relics treasured in the church experienced equal indignities. All the shrines became the prey of the eager avarice of the Huguenots; and the images of the saints and martyrs, torn from their tabernacles, graced the gibbets which were erected to receive them in various parts of Rouen.

Dom Pommeraye, in reciting these deplorable events, rises rather above his usual pitch of passion: "O malheur!" he exclaims, "ces corps sacrés, ces temples du Saint Esprit, qui avoient autrefois donné de la terreur aux Démons, ne trouverent ni crainte ni respect dans l'esprit de ces furieux, qui jetterent au feu tout ce qui tomba entre leurs mains impies et sacrilèges!" — The mischief thus occasioned was infinitely more to be lamented, he adds, than the burning of the church by the Normans;—"stones and bricks, and gold and jewels, may be replaced, but the loss of a relic is irreparable; and, moreover, the abbey thus forfeits a portion of its protection in heaven; for it is not to be doubted, but that the saints look down with eyes of peculiar favor upon the spots that contain their mortal remains; their glorified souls feeling a natural affection towards the bodies to which they are hereafter to be united for ever," on that day, when

> "Ciascun ritrovera la trista tomba,
>
> Ripigliera sua carne e sua figura,
>
> Udira ciò che in eterno rimbomba."

The outrages were curiously illustrative of the spirit of the times; the quantity of relics and ornaments equally characterise the devotion of the votaries, and the reputed sanctity of the place.

The royal abbey of St. Ouen had, indeed, enjoyed the veneration of the faithful, during a lengthened series of generations. Clothair is supposed to have been the founder of the monastery in 535; though other authorities claim for it a still higher degree of antiquity by one hundred and thirty years. The church, whoever the original founder may have been, was first dedicated to the twelve apostles; but, in 689, the body of St. Ouen was deposited in the edifice; miracles without number were performed at his tomb; pilgrims flocked thither; his fame diffused itself wider and wider; and at length, the allegiance of the abbey was tranferred to him whose sanctity gave him the best claims to the advocation.

Changes of this nature, and arising from the same cause, were frequent in those early ages: the abbey of St. Germain des Prés, at Paris, was originally dedicated to St. Vincent; that of Ste. Genevieve to St. Peter; and many other churches also took new patrons, as occasion required. According to one of the fathers of the church, the tombs of the beatified became the fortifications of the holy edifices: the saints were considered as proprietors of the places in which their bodies were interred, and where power was given them, to alter the established laws of nature, in favor of those who there implored their aid. But the aid which they afforded willingly to all their suitors, they could not bestow upon themselves. And oft, when the sword of the heathen menaced the land, the weary monks fled with the corpse of their patrons from the stubborn enemy. Thus, St. Ouen himself, on the invasion of the Normans, was transported to the priory of Gany, on the river Epte, and thence to Condé; but was afterwards conveyed to Rouen, when Rollo embraced Christianity. Other causes also contributed to the migration of these remains: they were often summoned in order to dignify acts of peculiar solemnity, or to be the witnesses to the oaths of princes, like the Stygian marsh of old,

"Dii cujus jurare timent et fallere numen."

William the Conqueror, upon the dedication of the abbey of St. Stephen, collected the bodies of all the saints in Normandy[92].

Those who wish to be informed of the acts and deeds of St. Ouen, may refer to Pommeraye's history of the convent, in which thirty-seven folio pages are filled with his life and miracles; the latter commencing while he was in long clothes. The monastery, under his protection, continued to increase in reputation; and, in the year 1042, the abbatial mitre devolved upon William, son of Richard IInd, Duke of Normandy, who laid the foundation of a new church, which, after about eighty years, was completed and consecrated by William Balot, next but one to him in the succession[93].

But this church did not exist long: ten years only had elapsed when a fire reduced it, together with the whole abbey, to ashes. An opportunity was thus afforded to the sovereign to shew his munificence, and Richard Cœur de Lion was not tardy in availing himself of it; but a second fire in 1248 again dislodged the monks; and they continued houseless, till the abbot, Jean Rousel, better known by the name of Mardargent, laid the foundation in 1318, of the present structure, an honor to himself, to the city, and to the nation. By this prelate the building was perfected as far as the transept: the rest was the work of subsequent periods, and was not completed till the prelacy of Bohier, who died in the beginning of the sixteenth century.

To speak more properly, I ought rather to say that it was not till then brought to its present state; for it was never completed. The western front is still imperfect. According to the original design, it was to have been flanked by magnificent towers, ending in a combination of open arches and tracery, corresponding with the outline and fashion of the central tower. These towers, which are now only raised to the height of about fifty feet, jut diagonally from the angles of the facade; and it was intended that, in the lower division, they should have been united by a porch of three arches, somewhat resembling the west entrance of Peterborough; and such as in this town is still seen, at St. Maclou, though on a much larger scale. Pommeraye has given an engraving of this intended front, taken from a drawing preserved in the archives of the abbey. The engraving is miserably executed; but it enables us to understand the lines of the projected building. Pommeraye has also preserved details of other parts of the church, among them of the beautiful rood-loft erected by the Cardinal d'Etouteville, and long an object of general admiration. The bronze doors of this screen were of a most singular and elegant pattern: Horace Walpole imitated them in his bed-room, at Strawberry-Hill. The rood-loft, which had been maimed by the Huguenots, was destroyed at the revolution; when the church was also deprived of its celebrated clock, which told the days of the month, the festivals, and the phases of the moon, and afforded other astronomical information. Such gazers as heeded not these mysteries, were amused by a little bronze statue of St. Michael, who sallied forth at every hour, and announced the progress of time, by the number of strokes which he inflicted on the Devil with his lance.

It is impossible to convey by words an adequate idea of the lightness, and purity, and boldness of St. Ouen. My imperfect description will be assisted by the sketches which I inclose. Of their merits I dare not speak; but I will warrant their fidelity; The flying buttresses end in richly crocketed pinnacles, supported by shafts of unusual height. The triple tiers of windows seem to have absorbed the solid wall-work of the building. Balustrades of varied quatrefoils run round the aisles and body; and the centre-tower, which is wholly composed of open arches and tracery, terminates, like the south-tower of the cathedral, with an octangular crown of fleurs-de-lys. The armorial symbol of France, which in itself is a form of great beauty, was often introduced by the French architects of the middle ages, amongst the ornaments of their edifices: it pleases the eye by its grace, and satisfies the mind by its appropriate and natural locality.

The elegance of the south porch is unrivalled. This portion of the church was always finished with care: it was the scene of many religious ceremonies, particularly of espousals. Hence they gave it a degree of magnitude which might appear disproportionate, did we not recollect that the arch was destined to embower the bride and the bridal train. The bold and lofty entrance of this porch is surrounded within by pendant trefoil arches, springing from carved bosses, and forming an open festoon of tracery. The vault within is ornamented with pendants, and the portal which it shades is covered with a profusion of sculpture: the death, entombment, and apotheosis of the Virgin, form the subjects of the principal groups. The sculptures, both in design and execution, far surpass any specimens of the corresponding æra in England. But this porch is now neglected and filled with lumber, and the open tracery is much injured. I hope, however, it will receive due attention; as the church is at this time under repair; and the restorations, as far as they go, have been executed with fidelity and judgment.

The perspective of the interior[94] is exceedingly impressive: the arches are of great height and fine proportions. If I must discover a defect, I should say that the lines appear to want substance; the mouldings of the arches are shallow. The building is all window. Were it made of cast iron, it could scarcely look less solid. This effect is particularly increased by the circumstance of the clerestory-gallery opening into the glazed tracery of the windows behind, the lines of the one corresponding with those of the other. To each of the clustered columns of the nave is attached a tabernacle, consisting of a canopy and pedestal, evidently intended originally to have received the image of a saint. It does not appear to have been the design of the architect that the pillars of the choir should have had similar ornaments; but upon one of them, at about mid-height, serving as a corbel to a truncated column, is a head of our Saviour, and, on the opposite pillar, one of the Virgin: the former is of a remarkably fine antique character. The capitals of the pillars in this part of the church were all gilt, and the spandrils of the arches painted with angels, now nearly effaced. The high altar is of grey marble, relieved, by a scarlet curtain behind, the effect of which is simple, singular, and good. Round the choir is a row of chapels, which are wholly wanting to the nave. The walls of these chapels have also been covered with fresco paintings; some with figures, others with foliage. The chapels contain many grave-stones displaying indented outlines of figures under canopies, and in other respects ornamented; but neglected, and greatly obliterated, and hastening fast to ruin. It is curious to see the heads and hands, and, in one instance, the crosier of a prelate, inlaid with white or grey marble; as if the parts of most importance were purposely made of the most perishable materials. I was much interested by observing, that many of these memorials are almost the exact counterparts of some of our richest English sepulchral brasses, and particularly of the two which are perhaps unrivalled, at Lynn[95].—How I wished that you, who so delight in these remains, and to whom we are indebted for the elucidation of those of Norfolk, had been with me, while I was trying to trace the resemblance

and particularly while I pored over the stone in the chapel of Saint Agnes, that commemorates Alexander Berneval, the master-mason of the building!

According to tradition, it was this same Alexander Berneval who executed the beautiful circular window in the southern transept. But being rivalled by his apprentice, who produced a more exquisite specimen of masonry in the northern transept, he murdered his luckless pupil. The crime he expiated with his own life; but the monks of the abbey, grateful for his labors, requested that his body might be entombed in their church; and on the stone that covers his remains, they caused him to be represented at full length, holding the window in his hand.

These large circular windows, sometimes known by the name of rose windows, and sometimes of marigold windows, are a strong characteristic feature of French ecclesiastical architecture. Few among the cathedrals or the great conventual churches, in this country, are without them. In our own they are seldom found: in no one of our cathedrals, excepting Exeter only, are they in the western front; and, though occasionally in the transepts, as at Canterbury, Chichester, Litchfield, Westminster, Lincoln and York, they are comparatively of small size with little variety of pattern. In St. Ouen, they are more than commonly beautiful. The northern one, the cause of death to the poor apprentice, exhibits in its centre the produced pentagon, or combination of triangles sometimes called the pentalpha.—The painted glass which fills the rose windows is gorgeous in its coloring, and gives the most splendid effect. The church preserves the whole of its original glazing. Each inter-mullion contains one whole-length figure, standing upon a diapered ground, good in design, though the artist seems to have avoided the employment of brilliant hues. The sober light harmonizes with the grey unsullied stone-work, and gives a most pleasing unity of tint to the receding arches.

Among the pictures, the-best are, the *Cardinal of Bologna opening the Holy Gate, instead of the Pope*, in the nave; and *Saint Elizabeth stopping the Pestilence*, in the choir: two others, in the Lady-Chapel, by an artist of Rouen, of the name of Deshays, the *Miracle of the Loaves*, and the *Visitation*, are also of considerable merit.—Deshays was a young man of great promise; but the hopes which had been entertained of him were disappointed by a premature death.

A church like this, so ancient, so renowned, and so holy, could not fail to enjoy peculiar privileges. The abbot had complete jurisdiction, as well temporal as spiritual, over the parish of St. Ouen; in the Norman parliament he took precedence of all other mitred abbots; by a bull of Pope Alexander IVth, he was allowed to wear the pontifical ornaments, mitre, ring, gloves, tunic, dalmatic, and sandals; and, what sounds strange to our Protestant ears, he had the right of preaching in public, and of causing the conventual bells to be rung whenever he thought proper. His monks headed the religious processions of the city; and every new archbishop of the province was not only consecrated in this church, but slept the evening prior to his installation at the abbey; whence, on the following day, he was conducted in pomp to the entrance of the cathedral, by the chapter of St. Ouen, headed by their abbot, who delivered him to the canons, with the following charge,—"Ego, Prior Sancti Audoeni, trado vobis Dominum Archiepiscopum Rothomagensem vivum, quem reddetis nobis mortuum."—The last sentence was also strictly fulfilled; the dean and chapter being bound to take the bodies of the deceased prelates to the church of St. Ouen, and restore them to the monks with, "Vos tradidistis nobis Dominum Archiepiscopum vivum; nos reddimus eum vobis mortuum, ita ut crastinâ die reddatis eum nobis."— The corpse remained there four and twenty hours, during which the monks performed the office of the dead with great solemnity. The canons were then compelled to bear the dead archbishop a second time from the abbey cross (now demolished) to the abbey of St. Amand[96], where the abbess took the pastoral ring from off his finger, replacing it by another of plain gold; and thence the bearers proceeded to the cathedral. These duties could not be very agreeable to portly, short-winded, well-fed dignitaries; and consequently the worthy canons were often inclined to shrink from the task. In the case of the funeral of Archbishop d'Aubigny, in 1719, they contented themselves with carrying him at once to his dormitory; but the prior and monks of St. Ouen instantly sued them before the parliament, and this tribunal decreed that the ancient service must be performed, and in default of compliance, the whole of their temporalities were to be put under sequestration: it is almost needless to add, that a sentence of excommunication would scarcely have been so effectual in enforcing the execution of the sentence.

The gardens formerly belonging to the abbey are at this time a pleasant promenade to the inhabitants of the town: the remains of the monastic buildings are converted into an Hôtel de Ville, where also the library and the museum are kept, and the academy hold their sittings. No remains, however, now exist of the abbatial residence, which was built by Anthony Bohier, in the beginning of the sixteenth century, and which, according to the engraving given of it by Pommeraye, must have been a noble specimen of domestic architecture. The sovereigns of France always took up their abode in it, during their visits to Rouen.—The circular tower called the Tour des Clercs, mentioned in a former letter, is the only vestige of Norman times.—The cloister corresponded with the architecture of the church: the south side of the quadrangle attached to the northern aisle still exists, but blocked up and dilapidated, and converted into a sort of cage for those who are guilty of disturbances during the night.

The church of St. Maclou is unquestionably superior to every other in the city, except the cathedral and St. Ouen. Its principal ornament are its

carved doors, produced during the reign of Henry IIIrd, by Jean Goujon, a man so eminent as to have been termed the Corregio of sculpture; but they have been materially injured by repairs and alterations by unskilful hands. Within the church, near the west entrance, is a singularly elegant stair-case, in filagree stone-work, which formerly led to the organ.—This building was erected in the year 1512, and chiefly by voluntary contributions, if such can be called voluntary as were purchased by promises from the archbishop, first of forty, and then of one hundred, days' indulgences, to all who would contribute towards the pious labor.—The central tower resembles that of the cathedral, both in the interior and the exterior. It now appears truncated; but it was originally surmounted by a spire, which was of such beauty, that even Italian artists thought it worthy to be engraved and held out as a model at Rome[97]. The spire, however, was greatly injured by a hurricane, in 1705, and it was at last taken down thirty years afterwards. To the triple porch, I have already alluded, in describing the intended front of St. Ouen. The general lines of the church, are such as in England would be referred to the fourteenth century: on a closer examination, however, the curious eye will discover the peculiar beauties of the French Gothic. Thus the bosses of the groined roof are wrought and perforated into filagree, the work extending over the intersections of the groins, which are seen through its reticulations. Such bosses are only found in the French churches of the sixteenth century. In other parts, the interior closely resembles the style of the cathedral[98].

St. Patricc is a building of the worst style of the commencement of the sixteenth century: to use the quaint phraseology of Horace Walpole, it exhibits "that *betweenity* which intervened when Gothic declined and Palladian was creeping in." The paintings on the walls of this church, and the stained glass in its windows, are more deserving of notice than its architecture. The first are of small size, and generally better than are seen in similar places. One of them is after Bassan, an artist, whose works are not often found in religious edifices in France. The painted windows of the choir deserve unqualified commendation. They are said to have been removed from St. Godard. Each is confined to a single subject; among which, that of the *Annunciation* is esteemed the best.

To this church was attached a confraternity[99], established in 1374, under the name of the Guild of the Passion. Its annual procession, which continued till the time of the revolution, took place on Holy-Thursday. It consisted of the usual pageantry; a host of children, dressed like angels, increased the train, which also included twelve poor men, whose feet the masters of the brotherhood publicly washed after mass. Like some other guilds, they were in possession of a pulpit or tribune, called, in old French, a Puy, from which they issued a general invitation to all poets, who were summoned to descant upon the themes which were commemorated by their union. The rewards held out to the successful candidates were, in the true monastic spirit of the guild, a reed, a crown of thorns, a sponge, or some other mystic or devotional emblem. Occasionally, too, they gave a scenic representation of certain portions of religious history, according to the practice of early times. The account of the Mystery of the Passion having been acted in the burial-ground of the church of St. Patrice, so recently as September, 1498, is preserved by Taillepied[100], who tells us, that it was performed by "bons joueurs et braves personages." The masters of this guild had the extraordinary privilege of being allowed to charge the expence attendant on the processions and exhibitions, upon any citizen they might think proper, whether a member or otherwise.

The neighboring church of St. Godard possesses neither architectural beauty, nor architectural antiquity; for, although it occupies the scite of an edifice of remote date, yet the present structure is coeval with St. Patrice. It has been supposed that this church was the primitive cathedral of the city[101]. One of the proofs of this assertion is found in a procession which, before the revolution, was annually made hither by the chapter of the present cathedral, with great ceremony, as if in recognition of its priority. The church was originally dedicated to the Virgin; but it changed its advocation in the year 525, when St. Godard, more properly called St. Gildard, was buried here in a subterranean chapel; and, for the reasons before noticed, the old tutelary patroness was compelled to yield to the new visitor. In the succeeding century, St. Romain, a saint of still greater fame, was also interred here; and, as I collect from Pommeraye[102], in the same crypt. This author strenuously denies the inferences which have been drawn from the annual procession, which he maintains was performed solely in praise and in honor of St. Romain; for the chapter, after having paid their devotions to the Host, descended into the chapel, to prostrate themselves before the sepulture of the saint; on which subject, an antiquary[103] of Rouen has preserved the following lines:—

"Ad regnum Domini dextrâ invitatus et ore,

Huic sacra Romanus credidit ossa loco;

Sontibus addixit quæ cæca rebellio flammis,

Nec tulit impietas majus in urbe scelus.

Quid tanto vesana malo profecit Erynnis?

Ipsa sui testis pignoris extat humus.

Crypta manet, memoresque trahit confessio cives,

Nec populi fallit marmor inane fidem.

Orphana, turba, veni, viduisque allabere saxis,

Est aliquid soboli patris habere thorum."

The body of St. Godard was carried to Soissons; but the tomb, which, has doubtfully been designated as appropriated either to him or to St. Romain, was left to the church, and remained there at least till the revolution. I have even been told that it is there still; but I had no opportunity of going down into the chapel to verify this point. It consisted, or rather consists, of a single slab of jasper, seven and a half feet long, by two feet wide, and two feet four inches thick. Upon it was this inscription:—

"Malades, voulez-vous soulager vos douleurs?

Visitez ce tombeau, baignez-le de vos pleurs;

Rechauffez vos esprits d'une divine flame;

Touchez-le settlement du doigt,

Et vous y trouverez (si vous avez la foi)

Et la santé du corps, et la santé de l'ame."

The building retains, at this time, only two of its celebrated painted windows; but they are fortunately the two which were always considered the best. One of them represents the history of St. Romain; the other, the genealogy of Jewish kings, from whom the Holy Virgin descended. Rouen has, from a very early period, been famous for its manufactories of painted glass. But the windows of this church were still esteemed the chef d'œuvre of its artists; and these had so far passed into a proverb, that Farin[104] tells us it was common throughout France to say, in recommendation of choice wine, that "it was as bright as the windows of St. Godard." The saying, however, was by no means confined to Rouen, for it was also applied to the windows of the Ste. Chapelle, at Dijon.

It was at St. Godard that the burst of the reformation was first manifested. The Huguenots, taking courage from the secret increase of their numbers, broke into the building, in 1540, demolished the images, and sold the pix to a goldsmith. But the man suffered severely for his purchase: he was shortly afterwards sentenced, by a decree of the parliament, to be hanged in front of his shop; and two of those concerned in the outrage also suffered capital punishment. The spark thus lighted, afterwards increased into a conflagration; and, to this hour, there is a larger body of Protestants at Rouen, than in most French towns.

I do not expect that you will reproach me with the prolixity of these details. The subject is attractive to me, and I feel that you will accompany me with pleasure in my pilgrimage, from chapel to shrine, dwelling with me in contemplation on the relics of ancient skill and the memorials of the piety of the departed. Nor must it be forgotten, that the hand of the spoliator is falling heavily on all objects of antiquity. And the French seem to find a source of perverse and malignant pleasure in destroying the temples where their ancestors once worshipped: many are swept away; a greater number continue to exist in a desecrated state; and time, which changes all things, is proceeding with hasty strides to obliterate their character. The lofty steeple hides its diminished head; the mullions and tracery disappear from the pointed windows, from which the stained glass has long since fallen; the arched entrance contracts into a modern door-way; the smooth plain walls betray neither niches, nor pinnacles, nor fresco paintings; and in the warehouse, or manufactory, or smithy, little else remains than the extraordinary size, to point out the original holy destination of the edifice.

Footnotes:

[91]   The following brief statement of their excesses is copied from a manuscript belonging to the monastery: the full detail of them engages Pommeraye for nearly seven folio pages:—"Le Dimanche troisiéme de May, 1562, les Huguenots s'étans amassez en grosse troupe, vinrent armez en grande furie dans l'Eglise de S. Ouen, où étant entrez ils rompirent les chaires du choeur, le grand autel, et toutes les chapelles: mirent en pieces l'Horloge, dont on voit encore la menuiserie dans la chapelle joignant l'arcade du costé du septentrion, aussi bien que celles des orgues, dont ils prirent l'étaim et le plomb pour en faire des balles de mousquet: puis ils allumerent cinq feux, trois dedans l'Eglise et deux dehors, où ils brûlerent tous les bancs et sieges des religieux, auec le bois des balustres des chapelles, les bancs et fermetures d'icelles, plusieurs ornemens et vestemens sacrez, comme chappes, tuniques, chasubles, aubes, vne autre partie des plus riches et precieux ornemens de broderie et drap d'or ayant esté enlevée en l'hôtellerie de la pomme de pin, où ils les brûlerent pour en auoir l'or et l'argent. Ils firent la mesme chose des saintes reliques, qu'ils brûlerent, ayant emporté l'or, l'argent, et les pierreries des reliquaires."—Histoire de l'Abbaye Royale de St. Ouen, p. 205.

[92] Farin, Histoire de Rouen, IV. p. 134.

[93] Histoire de l'Abbaye Royales de Saint Ouen, p. 204.

[94] The following are the dimensions of the interior of the building, in French feet:

Length of the church	416
Ditto of the nave	234
Ditto of the choir	108
Ditto of the Lady-Chapel	66
Ditto of the transept	130
Width of ditto	34
Ditto of nave, without the aisles	34
Ditto, including ditto	78
Height of roof	100
Ditto of tower	240

[95] Figured in Cotmans Norfolk Sepulchral Brasses.

[96] The house of the abbess of St. Amand is still standing, though neglected, and in a great degree in ruins. What remains, however, is very curious; and is, perhaps, the oldest specimen of domestic architecture in Rouen. It is partly of wood, the front covered with arches and other sculpture in bas-relief, and partly of stone.

[97] Farin, Histoire de Rouen, IV. p. 156.

[98] The dimensions of the building, in French feet, are, —

Length of the nave	70
Ditto of choir	40
Ditto of Lady-Chapel	30
Ditto of the whole building	140
Width of ditto	76
Height to the top of the lanthorn	142

[99] Farin, Histoire de Rouen, IV. p. 168.

[100] Antiquitéz et Singularitéz de la Ville de Rouen, p. 186.

[101] Farin, Histoire de Rouen, IV. p. 132.

[102] Histoire des Archevêques de Rouen, p. 130.

[103] La Normandie Chrétienne, p. 487.

[104] Histoire de Rouen, IV. p. 134.

LETTER XII
PALAIS DE JUSTICE—STATES, EXCHEQUER, AND PARLIAMENT OF NORMANDY—GUILD OF THE CONARDS—JOAN OF ARC—FOUNTAIN AND BAS-RELIEF IN THE PLACE DE LA PUCELLE—TOUR DE LA GROSSE HORLOGE—PUBLIC FOUNTAINS—RIVERS AUBETTE AND ROBEC—HOSPITALS—MINT

(Rouen, June, 1818.)

Amongst the secular buildings of Rouen, the Palais de Justice holds the chief place, whether we consider the magnificence of the building, or the importance of the assemblies which once were convened within its precinct.

The three estates of the Duchy of Normandy, the parliament, composed of the deputies of the church, the nobility, and the good towns, usually held their meetings in the Palace of Justice. Until the liberties of France were wholly extirpated by Richelieu, this body opposed a formidable resistance to the crown; and the Charte Normande was considered as great a safeguard to the liberties of the subject, as Magna Charta used to be on your side of the channel. Here, also, the Court of Exchequer held its session. According to a fond tradition, this, the supreme tribunal of Normandy, was instituted by Rollo, the good Duke, whose very name seemed to be considered as a charm averting violence and outrage. This court, like our Aula Regia, long continued ambulatory, and attendant upon the person of the sovereign; and its sessions were held occasionally, and at his pleasure. The progress of society, however, required that the supreme tribunal should become stationary and permanent, that the suitors might know when and where they might prefer their claims. Philip the Fair, therefore, about the year 1300, began by enacting that the pleas should be held only at Rouen. Louis the XIIth remodelled the court, and gave it permanence; yielding in these measures to the prayer of the States of Normandy, and to the advice of his minister, the Cardinal d'Amboise. It was then composed of four presidents, and twenty-eight counsellors; thirteen being clerks; and the remainder laymen. The name of exchequer was perhaps unpleasing to the crown, as it reminded the Normans of the ancient independence of their duchy; and, in 1515, Francis Ist ordered that the court should thenceforward be known as the Parliament of Normandy; thus assimilating it in its appellation to the other supreme tribunals of the kingdom. There is an old poem extant, written in very lawyer-like rhyme, which invests all the cardinal virtues, and a great many supernumerary ones besides, with the offices of this most honorable court, in which purity is the usher, truth has a silk gown, and virginity enters the proceedings on the record.

> "De ceste court grace est grand chanceliere,
>
> Vertus ont lieu de présidens prudens:
>
> Vérité est première conseillere,
>
> Et pureté huyssiére là-dedans:
>
> La greffiére est virginité féconde,
>
> Et la concierge humilité profonde.
>
> Pythié procure a vuider les discords,
>
> Comme advocat, amour ayde aux accords.
>
> De geolier vacque le seul office:
>
> Aussy on voyt par officiers concors,
>
> La noble court rendante à tous justice."

In the same style and strain is a ballad, which, thanks to the care of De Bourgueville, the author of the *Antiquities of Caen*, hath been preserved for the edification of posterity. It enumerates all the members of the court *seriatim*, and compares their lordships and worships, one after another, to the heroes and demi-gods of ancient story.

The parliament in its turn has given way to the *Court of Assizes*; and, where the states once deliberated, the electors of the department now come together for the purpose of naming the deputies who represent them in the great council of the nation;—such are the vicissitudes of all human institutions.

When the Jews were expelled from Normandy, in 1181, the *Close*, or Jewry, in which they dwelled, escheated to the king. The sons of Japhet spoiled the sons of Shem with pious alacrity. The debtor burnt his bond; the bailie seized the store of bezants; the synagogue was razed to the ground. In this *Close* the palace was afterwards built. The wise custom of Normandy was mooted on the spot where the law of Moses had once been taught; and, by a strange, perhaps an ominous, fatality, the judge held the scales of justice, where whilome the usurer had poised his balance.

The palace forms three sides of a quadrangle. The fourth is occupied by an embattled wall and an elaborate gate-way. The building was erected about the beginning of the sixteenth century; and, with all its faults, it is a fine adaptation of Gothic architecture to civil purposes. It is in the style which a friend of mine chooses to distinguish by the name of Burgundian architecture; and he tells me that he considers it as the parent of our Tudor style. Here, the windows in the body of the building take flattened elliptic heads; and they are divided by one mullion and one transom. The mouldings are highly wrought, and enriched with foliage. The lucarne windows are of a different design, and form the most characteristic feature of the front: they are pointed and enriched with mullions and tracery, and are placed within triple canopies of nearly the same form, flanked by square pillars, terminating in tall crocketed pinnacles, some of them fronted with open arches crowned with statues. The roof, as is usual in French and Flemish buildings of this date, is of a very high pitch, and harmonizes well with the proportions of the building. An oriel, or rather tower, of enriched workmanship projects into the court, and varies the elevations. On the left-hand side of the court, a wide flight of steps leads to the hall called la Salle des Procureurs, a place originally designed as an Exchange for the merchants of the city, who had previously been in the habit of assembling for that purpose in the cathedral. It is one hundred and sixty feet in length, by fifty in breadth.

"In this great hall," says Peter Heylin, "are the seats and desks of the procurators; every one's name written in capital letters over his head. These procurators are like our attornies; they prepare causes, and make them ready for the advocates. In this hall do suitors use, either to attend on, or to walk up and down, and confer with, their pleaders."—The attornies had similar seats in the ancient English courts of justice; and these seats still remain in the hall at Westminster, in which the Court of Exchequer holds its sittings. The walls of the Salle des Procureurs are adorned with chaste niches. The coved roof is of timber, plain and bold, and destitute either of the open tie-beams and arches, or the knot-work and cross timber which adorn our old English roofs. If the roof of our priory church was not ornamented, as last mentioned, it would nearly resemble that in question.—Below the hall is a prison; to its right is the room where the parliament formerly held its sittings, but which is now appropriated to the trial of criminal causes. The unfortunate Mathurin Bruneau, the soi-disant dauphin, was last year tried here, and condemned to imprisonment. He is treated in his place of confinement with ambiguous kindness. The poor wretch loves his bottle;

and, being allowed to intoxicate himself to his heart's content, he is already reduced to a state of idiotism.—Heylin, who saw the building when it was in perfection, says, speaking of this Great Chamber, "that it is so gallantly and richly built, that I must needs confess it surpasseth all the rooms that ever I saw in my life. The palace of the Louvre hath nothing in it comparable; the ceiling is all inlaid with gold, yet doth the workmanship exceed the matter."—The ceiling which excited Heylin's admiration still exists. It is a grand specimen of the interior decoration of the times. The oak, which age has rendered almost as dark as ebony, is divided into compartments, covered with rich but whimsical carving, and relieved with abundance of gold. Over the bench is a curious old picture, a Crucifixion. Joseph and the Virgin are standing by the cross: the figures are painted on a gold ground; the colors deep and rich; the drawing, particularly in the arms, indifferent; the expression of the faces good. It was upon this picture that witnesses took the oaths before the revolution; and it is the only one of the six formerly in this situation that escaped destruction[105]. Round the apartment are gnomic sentences in letters of gold, reminding judges, juries, witnesses, and suitors, of their duties. The room itself is said to be the most beautiful in France for its proportions and quantity of light. In the Antiquités Nationales, is described and figured an elaborately wrought chimney-piece in the council-chamber, now destroyed, as are some fine Gothic door-ways, which opened into the chamber. The ceiling of the apartment called la seconde Chambre des Enquêtes, painted by Jouvenet, with a representation of Jupiter hurling his thunderbolts at Vice, is also unfortunately no more. It fell in, from a failure in the woodwork of the roof, on the first of April, 1812. It was among the most highly-esteemed productions of this master, and not the less remarkable for having been executed with the left hand, after a paralytic stroke had deprived him of the use of the other.

Millin observes, with much justice, that one of the most remarkable of the decrees that issued from this palace, was that which authorized the meetings of the Conards, a name given to a confraternity of buffoons, who, disguised in grotesque dresses, performed farces in the streets on Shrove Tuesday and other holidays. Nor is it a little indicative of the taste of the times, that men of rank, character, and respectability entered into this society, the members of which, amounting to two thousand five hundred, elected from among themselves a president, whom they dressed as an abbot[106], with a crozier and mitre, and, placing him on a car drawn by four horses, led him, thus attired, in great pomp through the streets; the whole of the party being masked, and personating not only the allegorical characters of avarice, lust, &c. but the more tangible ones of pope, king, and emperor, and with them those of holy writ. The seat of this guild was at

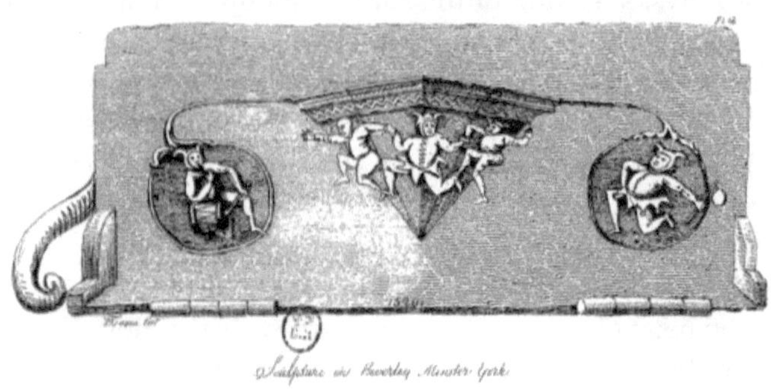

Notre Dame de Bonnes Nouvelles.

In the cathedral itself the more notorious Procession des Fous was also formerly celebrated, in which, as you know, the ass played the principal part, and the choir joined in the hymn[107], —

"Orientis partibus

Adventavit Asinus," &c.

These, or similar ceremonies, call them if you please absurdities, or call them impieties, (you will in neither case be far from their proper name,) were in the early ages of Christianity tolerated in almost every place. Mr. Douce has furnished us with some curious remarks upon them in the eleventh volume of the Archaeologia, and Mr. Ellis in his new edition of Brand's Popular Antiquities. I am indebted to the first of these gentlemen for the knowledge that the inclosed etching, copied some time ago from a drawing by Mr. Joseph Harding, is allusive to the ceremony of the feast of fools, and does not represent a group of morris-dancers, as I had erroneously supposed. Indeed, Mr. Douce believes that many of the strange carvings on the misereres in our cathedrals have references to these practices. And yet, to the honor of England, they never appear to have been equally common with us as in France. — According to Du Cange[108], the confraternity of the Conards or Cornards was confined to Rouen and Evreux. I have not been able to ascertain when they were suppressed; but they certainly existed in the time of Taillepied, in the beginning of the seventeenth century, about fifty years previously to which they dropped their original name of Coqueluchers. At this time too they had evidently degenerated from the primary object of their institution, "ridendo castigare mores atque in omne quod turpitèr factum fuerat ridiculum immittere." Taillepied was an eye-witness of their practices; and he prudently contents himself with saying; "le fait est plus clair à le voir que je ne pourrois icy l'escrire."

At a short distance from the palace is a small square, called the Place de la Pucelle, a name which it has but recently acquired, in lieu of the more familiar appellation of le Marché aux Veaux. The present title records one of the most interesting events in the history of Rouen, the execution of the unfortunate Joan of Arc, which is said to have taken place on the very spot now covered by the monument that commemorates her fate. Three different ones have in succession occupied this place. The first was a cross, erected in 1454, only twenty-four years after her death; for even at this early period, the King of France had obtained from Pope Calixtus IIIrd, a bull directing the revision of her sentence, and he had caused her innocence to be acknowledged. The second was a fountain of delicate workmanship, consisting of three tiers of columns placed one above the other, on a triangular plan, the whole decorated with arabesques and statues of saints, while the Maid herself crowned the summit, and the water flowed through pipes that terminated in horses' heads. The present monument is inferior to the second, equally in design and in workmanship: it is a plain triangular pedestal, ornamented with dolphins at the base, and surmounted by the heroine in military costume. Of the two last, figures are given by Millin[109], who could not be expected to suffer a subject to escape him, so calculated for the gratification of national pride. In a preceding volume of the same work[110], he has represented the monument erected to her memory by Charles VIIth, upon the bridge at Orleans: the latter is commemorative of her triumphs; that at Rouen, only of her capture and death. But the King testified his gratitude by more substantial tokens: he ennobled her three brothers and their descendants; and even allowed the females of the family to confer their rank upon the persons whom they married, a privilege which they continued to enjoy till the time of Louis XIIIth, who abolished it in 1634.

In the square is a house within a court, now occupied as a school for girls, of the same æra as the Palais de Justice, and in the same Burgundian style, but far richer in its sculptures. The entire front is divided into compartments by slender and lengthened buttresses and pilasters. The intervening spaces are filled with basso-relievos, evidently executed at one period, though by different masters. A banquet beneath a window in the first floor, is in a good cinque-cento style. Others of the basso-relievos, represent the labors of the field and the vineyard; rich and fanciful in their costume, but rather wooden in their design: the Salamander, the emblem of Francis Ist, appears several times amongst the ornaments, and very conspicuously. I believe there is not a single square foot of this extraordinary building, which has not been sculptured.—On the north side extends a spacious gallery. Here the architecture is rather in Holbein's manner: foliaged and swelling pilasters, like antique candelabra, bound the arched windows. Beneath, is the well-known series of bas-reliefs, executed on marble tablets, representing the interview between Francis Ist of France, and Henry VIIIth of England, in the Champ du Drap d'or, between Guisnes and Ardres. They were first discovered by the venerable father Montfaucon, who engraved them in his Monumens de la Monarchie Française[111]; but to the greater part of our antiquaries at home, they are, perhaps, more commonly known by the miserable copies inserted in Ducarel's work, who has borrowed most of his plates from the Benedictine.—These sculptures are much mutilated, and so obscured by smoke and dirt, that the details cannot be understood without great difficulty. The corresponding tablets above the windows, are even in a worse condition; and they appear to have been almost unintelligible in the time of Montfaucon, who conjectures that they were allegorical, and probably intended to represent the triumph of religion. Each tablet contains a triumphal car, drawn by different animals, one by elephants, another by lions, and so on, and crowded with mythological figures and attributes.—A friend of mine, who examined them this summer, tells me, that he thinks the subjects are either taken from the triumphs of Petrarch, or imitated from the triumphs introduced in the Polifilo. Graphic representations of allegories are susceptible of so many variations, that an artist, embodying the ideas of the poet, might produce a representation bearing a close resemblance to the mythological processions of the mystic dream.—Of one of the most perfect of the historical subjects, I send you a drawing: it is the first in order in Montfaucon's work, and exhibits the suite of the King of England, on their way from the town of Guisnes, to meet the French monarch. Two of the figures might be mistaken for Henry himself and Wolsey, riding familiarly

side by side; but these dignified personages have more important parts allotted them in the second and third compartments, where they appear in the full-blown honors of their respective characters.

The interior has been modernized; so that a beam covered with small carvings is the only remaining object of curiosity. On the top, a bunch of leaden thistles has been a sad puzzle to antiquaries, who would fain find some connection between the building and Scotland; but neither record nor tradition throw any light upon their researches. Montfaucon, copying from a manuscript written by the Abbé Noel, says, "I have more than once been told that Francis Ist, on his way through Rouen, lodged at this house; and it is most probable, that the bas-reliefs in question were made upon some of these occasions, to gratify the king by the representation of a festival, in which he particularly delighted." The gallery sculptures are very fine, and the upper tier is much in the style of Jean Goujon. It is not generally known that Goujon re-drew the embellishments of Beroald de Verville's translation of the Polifilo; and that these, beautiful as they are in the Aldine edition, acquired new graces from the French artist. — I have remarked that the allegorical tablets appear to coincide with the designs of the Polifilo: a more accurate examination might, perhaps, prove the fact; and then little doubt would remain. The building is much dilapidated; and, unless speedily repaired, these basso-relievos, which would adorn any museum, will utterly perish. In spite of neglect and degradations, the aspect of the mansion is still such that, as my friend observed, one would expect to see a fair and stately matron standing in the porch, attired in velvet, waiting to receive her lord. — In the adjoining house, once, probably, a part of the same, but now an inn, bearing the sign of la Pucelle, is shewn a circular room, much ornamented, with a handsome oriel conspicuous on the outside. In this apartment, the Maid is said to have been tried; but it is quite certain that not a stone of the building was then put of the quarry.

Hence I must take you, and still under the auspices of Millin[112], to the great town-clock, or, as it is here called, la Tour de la Grosse Horloge; and I cannot help wishing on the occasion, that I had half the powers of instructing and amusing which he possessed. Like the writers in our most popular Reviews, he uses the subjects which he places at the head of his articles as little more than a peg, whereon to hang whatever he knows connected with the matter; and the result is, that he is never read without pleasure or information. Such is peculiarly the case in the present instance, in which he takes an opportunity of giving the history of the origin of clocks, tracing them from the simple dial, and particularising the most curious and intricate contrivances of modern ingenuity. Another name of the tower which contains this clock, is la Tour du Beffroi, or, as we should say in English, the Belfry; for the two words have the same meaning, and it is not to be doubted but that they originated from the same root, the Anglo-Saxon bell, whence barbarous Latinists have formed Belfredus and

Berfredus, terms for moveable towers used in sieges, and so denominated from their resemblance in form to bell-towers. I mention this etymology, because the French have misled themselves strangely on the subject; and one of them has wandered so widely in his conjectures, as to derive beffroi from bis effroi, supposing it to be the cause of double alarm! Happily, in the most alarming of all times for France, that of the revolution, this bell, though appointed the tocsin, had scarcely ever occasion to sound. There is, however, another purpose, alarming at all periods, and especially in a town built of wood, to which it is appropriated, and to which we only yesterday heard it applied, the ringing to announce a fire. The precautions taken against similar accidents in Rouen, are excellent, and they had need be so; for insurance-companies of any kind are unknown, I believe, in France[113], or exist only upon a most limited scale, at the foot of the Pyrenees, where the farmers mutually insure each other against the effects of the hail. The daily office of this bell is to sound the curfew, a practice which, under different names, is still kept up through Normandy. Here it rings nightly at nine. In other towns it rings at nine in winter only, but not till ten in summer. In some places it is called la retraite.

Adjoining the bell-tower is a fountain, ornamented with statues of Alpheus and Arethusa, united by Cupid; a specimen of the taste of the far-famed siècles de Louis XIV et de Louis XV, and a worthy companion of the water-works at Versailles. There are in Rouen more than thirty public fountains, all supplied by five different springs, among which, those of Yonville and of Darnétal are accounted to afford the purest water.—The Robec and the Aubette also flow through Rouen in artificial channels. St. Louis granted them both to the city in 1262; but it was the great benefactor of the place, the Cardinal d'Amboise, who brought them within the walls, by means of a canal, which he caused to be dug at his own expence. For a space of two leagues their banks are uninterruptedly lined with mills and manufactories of various descriptions; and it is this circumstance which has given rise to the saying, that Rouen is a wonderful place, for "that it has a river with three hundred bridges, and whose waters change their color ten times a day."

As a building, the fountain of Lisieux, decorated with a bas-relief representing Parnassus, with Apollo, the Muses, and Pegasus, is most frequently pointed out to strangers; a wretched specimen of wretched taste. Infinitely more interesting to us are the Gothic fountains or conduits, which are now wholly wanting in England. Such is the fountain de la Croix de Pierre, which, in shape, style, and ornaments, resembles the monumental crosses erected by; our King Edward Ist, for his Queen Eleanor. The water flows from pipes in the basement. The stone statues, which filled the tabernacles, were destroyed during the revolution: they have been replaced by others in wood.—The fountain de la Crosse is of inferior size, and more recent date. It is a polygon, with sides of pannelled work, each compartment occupied by a pointed arch, with tracery in the spandrils. It ends in a short truncated pyramid, which, in Millin's time, was surmounted by a royal crown[114]. Its name is taken from a house, at whose corner it stands, and on whose roof was originally a crozier.

Writing to a friend may be regarded, if we extend to writing the happy comparison which Lord Bacon has applied to conversation, not as walking in a high-road which leads direct to a house, but rather as strolling through a country intersected with a variety of paths, in which the traveller wanders as fancy or accident directs. Hence I shall scarcely apologize for my abrupt transition to another very different subject, the hospitals.—There are at Rouen two such establishments, situated at opposite extremes of the town, the *Hospice Général* and the *Hôtel Dieu*, more commonly called *la Madeleine*. The latter is appropriated only to the sick; the former is also open to the aged, to foundlings, to paupers, and to lunatics. For the poor, I have been able to hear of no other provision; and poor-laws, as you know, have no existence in France; yet, even here, in a manufacturing town, and at a season of distress, beggary is far from extreme. These institutions, like all the rest at Rouen, are said to be under excellent management.

The annual expences of la Madeleine are estimated at two hundred and forty thousand-francs[115]; out of which sum, no less than forty-seven thousand francs are expended in bread. The number of individuals admitted here, during the first nine months of 1805, the last authentic statement I have been able to procure, was two thousand seven hundred and seventeen: during the same period, two thousand one hundred and fifty-eight were discharged, and two hundred and seventy died. The building is modern and handsome, and situated at the end of a fine avenue. The church, a Corinthian edifice, and indisputably the handsomest building of that description at Rouen, is generally admired. The Hospice Général, destitute as it is of architectural magnificence, cannot be visited without satisfaction. When I was at this hospital, the old men who are housed there were seated at their dinner, and I have seldom witnessed a more pleasing sight. They exhibited an appearance of cleanliness, propriety, good order, and comfort, equally creditable to themselves and to the institution. The number of inmates usually resident in this building is about two thousand; and they consisted, in 1805, of one hundred and sixty aged men, one hundred and eighty aged women, six hundred children, and eight hundred and twenty-five invalids. Among the latter were forty lunatics. The food here allowed to the helpless poor is of good quality; and, as far as I could learn, is afforded in sufficient quantity: there are also two work-shops; in one of which, articles are manufactured for the use of the house; in the other, for sale.

The principal towns of France, as was anciently the case in England, have each its mint. The numismatic antiquities of this kingdom are yet involved in considerable obscurity; but it is said that the monetary privileges of the towns were first settled by Charles the Bald[116], who, about the year 835, enacted, that money, which had previously only been coined in the royal palace itself, or in places where the sovereign was present, should be struck in future at Paris, Rouen, Rheims, Sens, Chalons sur Saone, Mesle in Poitou, and Narbonne. At present, the money struck at Rouen is impressed with the letter B, indicating that the mint is second only to that of Paris; for the city has remained in possession of the right of coinage throughout all its various changes of masters: it now holds it in common with ten other, cities in the kingdom. Ducarel[117] has figured two very scarce silver pennies, coined here by William the Conqueror, before the invasion of England; and Snelling and Ruding[118] detail ordinances for the regulation of the mintage of Rouen, during the reign of Henry Vth. I have not been able, however, to procure in the city any specimens of these, or of other Norman coins; and in fact the native spot of articles of virtu is seldom the place where they can be procured either genuine or in abundance. Greek medals, I am told, are regularly exported from Birmingham to Athens, for the supply of our travelled gentlemen; and, if groats and pennies should ever rise in the market, I doubt not but that they will find their way in plenty into the old towns of Normandy. There is not, at Rouen, any public collection of the productions of the mint. Since the annexation of the duchy to the crown of France, no coins have been struck here, except the common silver currency of the kingdom: the manufacture of medals and of gold coins is exclusively the privilege of the Parisian mint. The establishment is under the care of a commissary and assay-master, appointed by the crown, but not salaried. Their pay depends upon the amount of money coined, on which they are allowed one and a half per cent., and are left to find silver where they can; so that, in effect, it is little more than a private concern. The work is performed by four die-presses, moved by levers, each of which requires ten men; and about twenty thousand pieces can be produced daily from each press. But this method of working is attended with unequal pressure, and causes both trouble and uncertainty: it is even necessary that each coin should be separately weighed. The extreme superiority of the machinery of our own mint, where the whole operation is performed by steam, with a rapidity and accuracy altogether astonishing, affords Just reason for exultation to an Englishman.—It is true, that the execution of our bank paper rather counterbalances such feelings of complacency.

Footnotes:

[105] This appears from the following inscription now upon a silver tablet placed near it.—"Ce tableau est celui qui fut donné par Louis XII, en 1499, à l'Exchiquier, lorsqu'il le rendit permanent. C'est le seul de tous les ornemens de ce palais qui ait échappé aux ravages de la révolution: il a été conservé par les soins de M. Gouel, graveur, et par lui remis à la cour royale de Rouen qui l'a fait placer ici, comme un monument de la piété d'un roi, à qui sa bonté mérita le surnom de père du peuple, et dont les vertus se reproduisent aujourd'hui dans la personne non moins chérie que sacrée de sa majesté très chrétienne, Louis XVIII, 15 Janvier, 1816."

[106] Du Cange, (I. p. 24.) quoting from a book printed at Rouen, in 1587, under the title of Les Triomphes de l'Abbaye des Conards, &c. gives the following curious mock patent from the abbot of this confraternity, addressed to somebody of the name of De Montalinos.—

"Provisio Cardinalatus Rothomagensis Julianensis, &c.

"Paticherptissime Pater, &c.

"Abbas Conardorum et inconardorum ex quacumque Natione, vel genitatione sint aut fuerint: Dilecto nostro filio naturali et illegitimo Jacobo à Montalinasio salutem et sinistram benedictionem. Tua talis qualis vita et sancta reputatio cum bonis servitiis ... et quod diffidimus quòd postea facies secundùm indolem adolescentiæ ac sapientiæ tuæ in Conardicis actibus, induxenunt nos, &c. Quocirca mandamus ad amicos, inimicos et benefactores nostros qui ex hoc sæculo transierunt vel transituri sunt ... quatenus habeant te ponere, statuere, instalare et investire tàm in choro, chordis et organo, quàm in cymbalis bene sonantibus, faciantque te jocundari et ludere de libertatibus franchisiis, &c.... Voenundatum in tentorio nostro prope sanctum Julianum sub annulo peccatoris anno pontificatus nostri, 6. Kalend. fabacearum, hora verò noctis 17. more Conardorum computando, &c."

[107] The music of this hymn, or prose, as it is termed in the Catholic Rituals, is given in the Atlas to Millin's Travels through the Southern Departments of France, plate 4.

[108] See under the article Abbas Conardorum, I. p. 24.

[109] Antiquités Nationales, III. No. 36.

[110] Vol. II. No. 9.

[111] Vol. IV. t. 29, 30, 31.

[112] Antiquités Nationales, III. No. 30.

[113] This ceased to be the case almost immediately after this remark was made; for, on my return to France, in 1819, I observed on the whole road from Dieppe to Paris, the letters P A C I, or others, equally meaning pour assurance contre l'incendie, painted upon the fronts of the houses.

[114] Antiquités Nationales, III. article 30, p. 26.—(In the figure, however, which accompanies this article, the summit is mutilated, as I saw it.)

[115] Peuchet, Description Topographique et Statistique de la France, Département de la Seine Inférieure, p. 33.

[116] Histoire de la Haute Normandie, I. p. 94.

[117] Anglo-Norman Antiquities, p. 33. t. 3.

[118] Annals of the Coinage of Britain, I. p. 505-507.

LETTER XIII
MONASTIC INSTITUTIONS—LIBRARY—
MANUSCRIPTS—MUSEUM—ACADEMY—
BOTANIC GARDEN—THEATRE—
ANCIENT HISTORY—EMINENT MEN

(Rouen, June, 1818.)

The laws of France do not recognize monastic vows; but of late years, the clergy have made attempts to re-establish the communities which once characterized the Catholic church. To a certain degree they have succeeded: the spirit of religion is stronger than the law; and the spirit of contradiction, which teaches the subject to do whatever the law forbids, is stronger than either. Hence, most towns in France contain establishments, which may be considered either as the embers of expiring monachism, or the sparks of its reviving flame. Rouen has now a convent of Ursulines, who undertake the education of young females. The house is spacious; and for its neatness, as well as for the appearance of regularity and propriety, cannot be surpassed. On this account, it is often visited by strangers. The present lady-abbess, Dame Cousin, would do honor to the most flourishing days of the hierarchy: when she walks into the chapel, Saint Ethelburgha herself could not have carried the crozier with greater state; and, though she is somewhat short and somewhat thick, her pupils are all wonderfully edified by her dignity. She has upwards of dozen English heretics under her care; but she will not compromise her conscience by allowing them to attend the Protestant service. There are also about ninety French scholars, and the inborn antipathy between them and the insulaires, will sometimes evince itself. Amongst other specimens of girlish spite, the French fair-ones have divided the English damsels into two genera. Those who look plump and good-humored, they call Mesdemoiselles Rosbifs; whilst such as are thin and graver acquire the appellation of the Mesdemoiselles Goddams, a name by which we have been known in France, at least five centuries ago.— This story is not trivial, for it bespeaks the national feeling; and, although you may not care much about it, yet I am sure, that five centuries hence, it will be considered as of infinite importance by the antiquaries who are now babes unborn. The Ursulines and sœurs d'Ernemon, or de la Charité, who nurse the sick, are the only two orders which are now protected by government. They were even encouraged under the reign of Napoléon, who placed them under the care of his august parent, Madame Mère.—There are other sisterhoods at Rouen, though in small numbers, and not publickly patronized.

Nuns are thus increasing and multiplying, but monks and friars are looked upon with a more jealous eye; and I have not heard that any such communities have been allowed to re-assemble within the limits of the duchy, once so distinguished for their opulence, and, perhaps, for their piety and learning.

The libraries of the monasteries were wasted, dispersed, and destroyed, during the revolution; but the wrecks have since been collected in the principal towns; and thus originated the public library of Rouen, which now contains, as it is said, upwards of seventy thousand volumes. As may be anticipated, a great proportion of the works which it includes relate to theology and scholastic divinity; and the Bollandists present their formidable front of fifty-four ponderous folios.

The manuscripts, of which I understand there are full eight hundred, are of much greater value than the printed books. But they are at present unarranged and uncatalogued, though M. Licquet, the librarian, has been for some time past laboring to bring them into order. Among those pointed out to us, none interested me so much as an original autograph; of the *Historica Normannorum*, by William de Jumiegies, brought from the very abbey to which he belonged. There is no doubt, I believe, of its antiquity;

but, to enable you to form your own judgment upon the subject, I send you a tracing of the first paragraph.

I also add a fac-simile of the initial letter of the foregoing epistle, illuminated by the monk, and in which he has introduced himself in the act of humbly presenting his work to his royal namesake. I am mistaken, if any equally early, and equally well authenticated representation of a King of England be in existence. The Historia Normannorum is incomplete, both at the beginning and end, and it does not occupy more than one-fifth of the volume: the rest is filled with a comment upon the Jewish History.

The articles among the manuscripts, most valued by antiquaries, are a *Benedictionary* and a *Missal*, both supposed of nearly the same date, the beginning of the twelfth century.

The Abbé Saas, who published, in 1746, a catalogue of the manuscripts belonging to the library of the cathedral of Rouen, calls this Benedictionary, which then belonged to the metropolitan church, a Penitential; and gives it as his opinion, that it is a production of the eighth century, with which æra he says that the character of the writing wholly accords. Montfaucon, who never saw it, follows the Abbé; but the opinion of these learned men has recently been confuted by M. Gourdin[119], who has bestowed considerable pains upon the elucidation of the history and contents of this curious relic. He states that a sum of fifteen thousand francs had been offered for it, by a countryman of our own; but I should not hesitate to class this tale among the numberless idle reports which are current upon the continent, respecting the riches and the folly of English travellers. The famous Bedford Missal, at a time when the bibliomania was at its height[120], could hardly fetch a larger sum; and this of Rouen is in no point of view, except antiquity, to be put in competition with the English manuscript. Its illuminations are certainly beautiful; but they are equalled by many hundreds of similar works; and they are only three in number, the Resurrection, the Descent of the Holy Ghost, and the Death of the Virgin.—The volume appears to have been originally designed for the use of the cathedral of Canterbury; as it contains the service used at the consecration of our Anglo-Saxon sovereigns.

The Missal, which is also the object of M. Gourdin's dissertation, is from the convent of Jumieges. Its date is established by the circumstance of the paschal table finishing with the year 1095. It contains eleven miniatures, inferior in execution to those in the Benedictionary; and it ends with the following anathema, in the hand-writing of the Abbot Robert, by whom it was given to the monastery:—"Quem si quis vi vel dolo seu quoque modo isti loco subtraxerit, animæ suæ propter quod fecerit detrimentum patiatur, atque de libro viventium deleatur et cum justis non scribatur."

As a memorial of a usage almost universal in the earlier ages of the church, the Diptych, commonly called the Livre d'Ivoire, is a valuable relic. The covers exhibit figures of St. Peter and of some other saint, in a good style of workmanship, perhaps of the lower empire. The book contains the oaths administered to each archbishop of Rouen and his suffragans, upon their entering on their office, all of them severally subscribed by the individuals by whom they were sworn. It begins at a very early period, and finishes with the name of Julius Basilius Ferronde de la Ferronaye, consecrated Bishop of Lisieux, in 1784. In the first page is the formula of the oath of the archbishop.—"Juramentum Archiepiscopi Rothomagensis jucundo adventu receptionis suæ.—Primo dicat et pronuntiet Decanus vel alius de Majoribus verba quæ sequentur in introitu atrii;—Adest, reverende pater, tua sponsa, nostra mater, hæc Rothom. ecclesia, cum maximo gaudio recipere te parata, ut eam regas salubriter, potenter protegas et defendas.— Responsio Archiepiscopalis;—Hæc, Deo donante, me facturum promitto.— Iterum Decanus vel alius;—Firma juramento quæ te facturum promittis.— Ego, Dei patientia, bujus Rothom. ecclesiæ minister, juro ad hæc sancta Dei evangelia quod ipsam ecclesiam contra quoslibet tam in bona quam in personas ipsius invasores et oppressores pro posse protegam viriliter et defendam, atque etiam ipsius ecclesiæ jura, libertates, privilegia, statuta et consuetudines apostolicas servabo fideliter. Bona ejusdem ecclesiæ non alienabo nec alienari permittam, quin pro posse, si quæ alienata fuerint, revocabo. Sic me Deus adjuvet et sancta Dei evangelia."

The oath of the bishops and abbots was nothing more than a promise of constant respect and obedience on their parts to the church and archbishop of Rouen. You will find it in the Voyages Liturgiques[121]; in which you will also meet with a great deal of curious matter touching the peculiar customs and ceremonies of this cathedral. The different metropolitan churches of France before the revolution, like those of our own country prior to the reformation, varied materially from one another in observances of minor importance; at the same time that their rituals all agreed in what may be termed the doctrinal ceremonies of the church.

The last manuscript which I shall mention, is the only one that is commonly shewn to strangers: it is a Graduel, a very large folio volume, written in the seventeenth century, and of transcendent beauty. Julio Clovio himself, the Raphael of this department of art, might have been proud to be considered the author of the miniatures in it. The representations of lapis lazuli are even more wonderful than the flowers and insects. The whole was done by a monk, of the name of Daniel D'Eaubonne, and is said to have cost him the labor of his entire life.

In earlier times, a similar occupation was regarded as peculiarly meritorious[122].—There died a friar, a man of irregular life, and his soul was brought before the judgment-seat to receive its deserts. The evil spirits attended, not anticipating any opposition to the claim which they preferred; but the guardian angels produced a large book, filled with a transcript from holy writ by the hand of the criminal; and it was at length agreed that each letter in it should be allowed to stand against a sin. The tale was carefully gone through: Satan exerted his utmost ingenuity to substantiate every crime of omission or commission; and the contending parties kept equal pace, even unto the last letter of the last word of the last line of the last page, when, happily for the monk, the recollection of his accuser failed, and not a single charge could be found to be placed in the balance against it. His soul was therefore again remanded to the body, and a farther time was allotted to it to correct its evil ways.—The legend is pointed by an apposite moral; for the brethren are exhorted to "pray, read, sing, and write, always bearing in mind, that one devil only is allowed to assail a monk who is intent upon his duties, but that a thousand are let loose to lead the idle into temptation."

The library is open every day, except Sundays and Thursdays, from ten to two, to everybody who chooses to enter. It is to the credit of the inhabitants of Rouen, that they avail themselves of the privilege; and the room usually contains a respectable assemblage of persons of all classes. The revenue of the library does not amount to more than three thousand francs per annum; but it is also occasionally assisted by government. The French ministers of state consider that it is the interest of the nation to promote the publication of splendid works, either by pecuniary grants to the authors, or, as more commonly happens, by subscribing for a number of copies, which they distribute amongst the public libraries of the kingdom.—I could say a great deal upon the difference in the conduct of the governments of France and England in this respect, but it would be out of place; and I trust that our House of Commons will not be long before they expunge from the statute-books, a law which, under the shameless pretence of "encouraging learning," is in fact a disgrace to the country.

The museum is also established at the Hôtel-de-Ville, where it occupies a long gallery and a room adjoining. It is under the superintendence of M. Descamps, son of the author of two very useful works, La Vie des Peintres Flamands and Le Voyage Pittoresque. The father was born at Dunkirk, in 1714, but lived principally at Paris, till an accidental circumstance fixed him at Rouen, in 1740. On his way to England, he here formed an acquaintance with M. de Cideville, the friend of Voltaire, who, anxious for the honor of his native town, persuaded the young artist to select it as the place of his future residence. The event fully answered his expectation; for the ability and zeal of M. Descamps soon gave new life to the arts at Rouen. A public academy of painting was formed under his auspices, to which he afforded gratuitous instruction; and its celebrity increased so rapidly, that the number of pupils soon amounted to three hundred; and Norman authors continued to anticipate in fancy the creation of a Norman school, which should rival those of Bologna and Florence, until the very moment when the revolution dispelled this day-dream. Descamps died at the close of the last century. To his son, who inherits his parent's taste, with no small portion of his talent, we were indebted for much obliging attention.

The museum is open to the public on Sundays and Thursdays; but daily to students and strangers. It contains upwards of two hundred and thirty paintings. Of these, the great mass is undoubtedly by French artists, comparatively little known and of small merit, imitators of Poussin and Le Brun. Such paintings as bear the names of the old Italian masters, are in general copies; some of them, indeed, not bad imitations. Among them is one of the celebrated Raphael, commonly called the *Madonna di San Sisto*, a very beautiful copy, especially in the head of the virgin, and the female saint on her left hand. It is esteemed one of his finest pieces; but few of his pictures are less generally known: there is no engraving of it in Landon's eight volumes of his works.

Looking to the unquestionable originals in the collection, there are perhaps none of greater value than Jouvenet's finished sketches for the dome of the Hôtel des Invalides, at Paris. They represent the twelve apostles, each with his symbol, and are extremely well composed, with a bold system of light and shadow. The museum has five other pictures by the same master;

in this number are his own portrait, a vigorous performance, as well in point of character as of color; and the Death of St. Francis, which has generally been considered one of his happiest works. Both these were painted with his left hand. The death of St. Francis is said to have been his first attempt at using the brush, after he was affected with paralysis, and to have been done by way of model for his scholar, Restout, whom he had desired to execute the same subject for him. A Christ bearing his Cross, by Polemburg; is a little piece of high finish and considerable merit; an Ecce Homo, by Mignard, is excellent; and a St. Francis in Extasy, by Annibal Caracci, is a good illustration of the true character of the Bolognese school: it is a fine and dignified picture, depending for its excellence upon a grand character of expression and drawing, and light and shade, and not at all on bright or varied coloring, to which it makes no pretension.

As local curiosities, the attention of the amateur should be devoted to the productions of the painters to whom Rouen has given birth, Restout, Lemonnier, Deshays, Leger, Houel, Letellier, and Sacquespée, artists, not of the first class, but of sufficient merit to do great credit to the exhibition of a provincial metropolis.

From these recent specimens, you would turn with the more pleasure to a picture by Van Eyck, the inventor, as it is generally supposed, of oil painting. Let us respect these fathers of the art. Let us pardon the stiffness of their composition, the formality of their figures, the inelegance of their draperies, the hardness of their outlines, and the want of chiaroscuro;—for, in spite of all these failings, there is a truth to nature, and a richness of coloring, which always attract and win. The picture in question is the Virgin Mother in her Domestic Retirement, surrounded by her family, a comely party of young females in splendid attire, some of them wearing the bridal crown. It is altogether a curiosity, partaking, indeed, of the general bad taste of the times, but painted with great attention to nature in the minutiæ, and resembling Lionardo da Vinci in many particulars, especially in the high finishing, the coloring of the carnations, and the grace, and beauty of some of the heads. The draperies, too, are rich and brilliant.

This museum is a recent erection: most, if not all, of the departments of France, possess similar establishments in their principal towns. The basis of the collection is founded upon the plunder of the suppressed monasteries; but M. Descamps told us that, in the course of a journey to Italy, he had been the means of adding to this, at Rouen, its principal ornaments. He had the greater merit of preserving it entire, when orders were transmitted from Paris to send off its best pictures, to replace those taken from the Louvre by the allies; for on all occasions, whether great or small, the interests of the departments are sacrificed without mercy to the engulphing capital. Descamps was firm in defending his trust: he resisted the spoliation, upon the principle that the museum was the private property of the town; and the plea was admitted.

The same conventual buildings also contain the rooms appropriated to the use of the academy at Rouen, a royal institution of old standing, and which has published fifteen volumes of its transactions.—It was founded in 1744, under a charter granted to the Duke of Luxembourg, then governor of the province, and its first president. The present complement of members consists of forty-six fellows, besides non-resident associates. Its meetings are held every Friday evening, and the members, as at the institute at Paris, read their own papers. A few nights ago, at a meeting of this academy, I heard a memoir from the pen of the professor of botany, in which he dwelt at large upon the family of the lilies, but prized and praised them for nothing so much as for their connection with the Bourbon family. I mention the fact to shew you how readily the French seize hold of every occasion of displaying their devotion to the powers that be. In 1814, at the moment of the restoration of Louis XVIIIth, we were not surprised to see every town and village between Calais and Paris, decorated with a proud display of the busts of the monarch, the shields of France and Navarre, and innumerable devices and mottoes, consecrated, as the French say, to the Bourbons; but four years have given time for this ebullition of loyalty to subside; and the introduction of such topics at the present day, and especially in the meetings of a body devoted solely to the improvement of literature and of the arts and sciences, appears to savor somewhat of adulation. These praises excited no remarks and no criticisms; though both might have been expected; for, during the reading of a paper, the by-standers are allowed to discuss its merits and its defects. This practice gives the sittings of a French literary society a degree of life and spirit wanting to ours in England; but I doubt if the advantage be not more than counter-balanced by the frequent interruptions which it occasions, and which an ill-natured person might in some cases suspect to proceed from a desire of attracting notice, rather than from fair, and just reprehension. I should be sorry to insinuate that any thing of this kind was evident at the time, just alluded to, which was the Friday previous to the annual meeting, the day appointed for taking into consideration the report intended to be submitted to the full assembly of the inhabitants. The president also read his projected speech, in the course of which he took the opportunity of declaring in strong terms his dislike to Napoléon's plan of education, directed almost exclusively to military affairs and mathematics: he even stated that the present generation "étoit sans morale."—The opinion could not be allowed to pass: he found himself beset on all sides; not an individual supported him; and after a variety of attempts to palliate and explain away the offensive passage, he was obliged to consent to expunge it. This will give some farther idea of the state of public feeling in France: the compliment upon the lilies passed as words of course; but the same body that tolerated it, positively refused to stamp with the sanction of their approbation, any comparison unfavorable to the system of Napoléon, when put in opposition to that of the subsisting government.

There is another literary body at Rouen; called la Société d'Emulation, of more recent establishment, it having been founded in 1791. Conformably to the national spirit which then prevailed, it is directed exclusively to the encouragement of manufactories and agriculture.—This society distributes annual medals as the reward of improvements and discoveries, though I am afraid that as yet it has been productive but of slender utility.

Rouen also possesses a Botanic Garden, which was founded in 1738; but the scite which it now occupies was not thus applied till twenty years subsequently, when the municipality conveyed the ground in perpetuity to the academy in its corporate capacity, stipulating that it should yield a nosegay every year as an appropriate *rent in kind*. At the revolution a grant like this would scarcely be respected; still less did the jacobins appreciate the pleasures or advantages derived from the garden. The demagogues of that period seem to have entered heartily into Jean Jacques Rousseau's notions, that the arts and sciences were injurious to mankind: this fine establishment was seized as national property, and, according to the revolutionary jargon, was *soumissioné*; but a more temperate faction obtained the ascendancy before the sale was carried into effect.—The collection is extensive, and the plants are in good order: I am not however, aware that the city has ever given birth to any man of eminence in this department of science. Lately, indeed, the Abbé Le Turquier Deslongchamps, a very well-informed botanist, as well as a most excellent man, has published a *Flore des Environs de Rouen*, in two volumes; and there are many instances in which such works have been known to diffuse a taste, which public gardens and the lectures of professors had in vain endeavored to excite.

The variety of soil in the vicinity of the city renders it eminently favorable to the study of botany. It is peculiarly rich in the Orchideæ of the most beautiful and interesting families of the vegetable kingdom. The curious Satyrium hircinun is found in the utmost profusion upon the chalky hills immediately adjoining the city; and, at but a few miles distance, in a continuation of the same ridge, the bare chalk, under the romantic hill of

St. Adrien, is purpled with the flowers of the Viola Rothomagensis, a plant scarcely known to exist in any other place.

The suburbs of Rouen abound with nursery-grounds and gardens: the former contribute greatly to the preservation of the genuine stock of apple-trees, which furnish the cider, for which Normandy has for many centuries been celebrated; the latter supply the inhabitants with the flowers which are seen at almost every window. The square in front of the cathedral is the principal flower-market; and the bloom and luxuriance and variety of the plants exposed for sale, render it a most pleasing promenade. Various species of jessamines and roses, with oleanders, pomegranates, myrtles, egg-plants, orange and lemon trees, the *Lilium superbum* and *tigrinum*, *Canna Indica*, *Gladiolus cardinalis*, *Clerodendrum fragrans*, *Datura ceratocolla*, *Clethra alnifolia*, and *Dianthus Carthusianorum*, are to be seen in the greatest profusion and beauty. They at once attest the care of the cultivators, and a climate more genial than ours. None of the flowers, however, excited my envy so much as the *Rosa moschata*, which grows here in the open air, and diffuses its delicious fragrance from almost every window of the town.

It is perhaps to the credit of Rouen, that science and learning appear to flourish more kindly than the drama. The theatre of Rouen is quite uncharacteristic of the passion which the French usually entertain for spectacles. The house is shabby; the audience, as often as we have been there, has been small; and in this great city, the capital of an extensive, populous, and wealthy district we have witnessed acting so wretched, as would disgrace the floor of a village barn. We have been much surprised by seeing the performers repeatedly laugh in the face of the spectators, a thing which I should least of all have expected in France, where usually, in similar cases, the whole nation is tremblingly alive to the slightest violations of decorum. And yet Corneille, the father of the French drama, was born in this city: the scene that is used for a curtain at the theatre bears his portrait, with the inscription, "P. Corneille, natif de Rouen;" and his apotheosis is painted upon the cieling. These recollections ought to tend to the improvement of the drama. The portrait of the great tragedian is more appropriate than the busts of Henry IVth and Louis XVIIIth, which occupy opposite sides of the stage; the latter laurelled and flanked with small white flags, whose staffs terminate in paper lilies.

Corneille and Fontenelle are the citizens, of whom Rouen is most proud: the house in which Corneille was born, in the Rue de la Pie, is still shewn to strangers. His bust adorns the entrance, together with an inscription to his honor. The residence of his illustrious nephew, the author of the Plurality of Worlds, is situated in the Rue des bans Enfans, and is distinguished in the same manner. The whole Siécle de Louis XIV, scarcely contains two names upon which Voltaire dwells with more pleasure.—Rouen was also the birth-place of the learned Bochart, author of Sacred Geography and of the Hierozöicon; of Basnage, who wrote the History of the Bible; of Sanadon, the translator of Horace; of Pradon, "damn'd," in the Satires of Boileau, "to everlasting fame;" of Du Moustier, to whom we are indebted for the Neustria Pia; of Jouvenet, whom I have already mentioned as one of the most distinguished painters of the French school; and of Father Daniel, not less eminent as an historian.—These, and many others, are gone; but the reflection of their glory still plays upon the walls of the city, which was bright, while they lived, with its lustre;—"nam præclara facies, magnæ divitiæ, ad hoc vis corporis, alia hujuscemodi omnia, brevi dilabuntur; at ingenii egregia facinora, sicuti anima, immortalia sunt. Postremò corporis et fortunæ bonorum, ut initium, finis est; omnia orta occidunt et aucta senescunt: animus incorruptas, æternus, rector humani generis, agit atque habet cuncta, neque ipse habetur."

The more remote and historical honors of Rouen would present ample materials. Prior to the Roman invasion, it appears to have been of less note than as the capital of Neustria.

Julius Cæsar, copious as he is in all that relates to Gaul, makes no mention of Rouen in his Commentaries. Ptolemy first speaks of it as the capital of the Velocasses, or Bellocasses, the people of the present Vexin; but he does not allow his readers to entertain an elevated idea of its consequence; for he immediately adds, that the inhabitants of the Pays de Caux were, singly, equal to the Velocasses and Veromandui together; and that the united forces of the two latter tribes did not amount to one-tenth part of those which were kept on foot by the Bellovaci.—Not long after, however, when the Romans became undisputed masters of Gaul, we find Rouen the capital of the province, called the Secunda Lugdunensis; and from that tine forward, it continued to increase in importance. Etymologists have been amused and puzzled by "Rothomagus," its classical name. In an uncritical age, it was contended that the name afforded good proof of the city having been founded by Magus, son of Samothes, contemporary of Nimrod. Others, with equal diligence, sought the root of Rothomagus in the name of Roth, who is said to have been its tutelary god; and the ancient clergy adopted the tradition, in the hymn, which forms a part of the service appointed for the feast of St. Mellonus,—

"Extirpate Roth idolo,

 Fides est in lumine;

 Ferro cinctus, pane solo

 Pascitur et flumine,

 Post hæc junctus est in polo

 Cum sanctorum agmine."

The partizans of *Roth* are therefore supported by the authority of the church; the favorers of *Magus* must defend themselves by more worldly erudition; and we must leave the task of deciding between the claims of the two sections of the word, divided as they are by the neutral *o*, to wiser heads than ours.

Footnotes:

[119]   Précis Analytique des travaux de l'Académie de Rouen, pendant l'année 1812, p. 164.

[120] At the sale of Mr. Edwards' library, in April 1815, it was bought by the present Duke of Marlborough for six hundred and eighty-seven pounds fifteen shillings.—The following anecdote, connected with it, was communicated to me by a literary friend, who had it from one of the parties interested; and I take this opportunity of inserting it, as worthy of a place in some future Bibliographical Decameron.—At the time when the Bedford Missal was on sale, with the rest of the Duchess of Portland's collection, the late King sent for his bookseller, and expressed his intention to become the purchaser. The bookseller ventured to submit to his Majesty, that the article in question, as one highly curious, was likely to fetch a high price.—"How high?"—"Probably, two hundred guineas!"—"Two hundred guineas for a Missal!" exclaimed the Queen, who was present, and lifted up her hands with extreme astonishment.—"Well, well," said his Majesty, "I'll still have it; but, since the Queen thinks two hundred guineas so enormous a sum for a Missal, I'll go no farther."—The bidding for the royal library did actually stop at that point; and Mr. Edwwards carried off the prize by adding three pounds more.

[121] Published at Rouen, A.D. 1718.—The book professes to be written by the Sieur de Moléon; but its real author was

Jean Baptiste de Brun Desmarets, son of a bookseller in that city.—He was born in 1650, and received his education at the Monastery of Port Royal des Champs, with the monks of which order he kept up such a connection, that he was finally involved in their ruin. His papers were seized; and he was himself committed to the Bastille, and imprisoned there five years. He died at Orleans, 1731.

[122] Ordericus Vitalis, in Duchesne's Scriptores Normanni, p. 470.

INDEX

—stone bridge built by Matilda,

—boulevards,

—grand cours,

—costume of the inhabitants,

—house-rent,

—annual expences of the city,

—population,

—probably a Roman station,

—old castles,

—halles,

—privilege of St. Romain,

—capitulation to Henry Vth,

—Château du Vieux Palais,

—petit Château,

—fort on Mont Ste. Catherine,

—priory upon ditto,

—taken by Charles IXth,

—mineral springs,

—church of St. Paul,

—church of St. Gervais,

—palace on the Mont aux Malades,

—old part of the church of St. Ouen,

—cathedral,

—church of St. Ouen,

—church of St; Maclou,

—church of St. Patrice,

—church of St. Godard,

—house of the Abbess of St. Amand,

—Palais de Justice,